THE CONTROL OF INDUSTRIAL PROCESSES

BY DIGITAL TECHNIQUES

THE CONTROL
OF INDUSTRIAL PROCESSES
BY DIGITAL TECHNIQUES

The Organisation, Design and Construction of Digital Control Systems

W S Blaschke

and

J McGill

ELSEVIER SCIENTIFIC PUBLISHING COMPANY

AMSTERDAM - OXFORD - NEW YORK   1976

ELSEVIER SCIENTIFIC PUBLISHING COMPANY
335 Jan van Galenstraat
P.O. Box 211, Amsterdam, The Netherlands

Distributors for the United States and Canada:

ELSEVIER/NORTH-HOLLAND INC.
52, Vanderbilt Avenue
New York, N.Y. 10017

Library of Congress Cataloging in Publication Data

Blaschke, W   S
    The control of industrial processes by digital
techniques.

    Bibliography: p.
    Includes index.
    1. Process control. 2. Digital control systems.
I. McGill, J., joint author. II. Title.
TS156.8.B57   621.7'028'54   76-41744
ISBN 0-444-41493-2

Printed in The Netherlands

PREFACE

In the manufacturing industry there are fewer examples of integrated control systems than the current state of control technology would suggest feasible. There are many reasons for this. One obstacle to the successful introduction of control systems is the difficulty of reaching agreement between the two main parties concerned. On one side there is the production engineer. He must understand what is required from the control operation in every detail. On the other side there is the control engineer. He has the task of meeting these requirements within the limitations of the techniques available to him. On each side other specialists become involved. If misunderstandings come to light too late in the implementation, the consequences can be disastrous.

In this book there is an attempt at bridging the gap by concentrating on the logical formulation of control requirements, before describing the relevant control techniques. However, under the same cover, many practical aspects of control technology are discussed (notably in Chapter 4). Even the man who conceives the control system must be aware of them, lest his demands turn out to be impractical.

In putting together such a document, the authors are aware of the criticism which can be levelled at them from both sides. The specialist in control engineering may decide that the treatment of the techniques is too elementary. The production engineer may consider the approach too theoretical and general, not giving enough precise points of guidance. The real difficulty is to give a treatment of the subject which is helpful in the great variety of situations which occur in practice.

The book is neither an introduction to the subject of digital control, nor a treatise on the state of the art in the relevant technologies. Its aim is to further the interdisciplinary approach to the subject. Therefore it has both to introduce subjects that are outside the specialist's territory and take them through to a point where he can take advantage of the latest techniques available. It should also help him to counteract biased views of the power of certain techniques.

The authors have come to believe that to every control problem there is a 'natural' solution. Advances in technology do not substantially change this; only the details of the implementation are affected. The starting point should always be a thorough analysis of the tasks of the control system. This will lead logically to the most appropriate form of organisation. It

will decide the balance of centralisation versus decentralisation, generality
versus particularity. Finally, there is the choice (governed by scientific
and technical, and to some extent cost, considerations) of the most appropriate
control mechanism for various parts of the system. Never should this reasoning
be reversed. A great deal of sterile controversy about the merits and demerits
of particular techniques can thereby be avoided.

The book is more of a compendium of experience gained than a tutorial
text. Few users would want to read it from cover to cover. However, they
might find something at odd points which really helps them over a particular
difficulty. The authors will then be satisfied that their efforts have been
rewarded.

As always, recognition is due to the innumerable colleagues with whom
one has exchanged views on various subjects and obtained clarification as a
result.

Particular acknowledgement is due to Michael Cook, Ann Ballantyne and
Isobel Boyd for assistance with editing, typing and the onerous task of
camera-ready preparation.

W S Blaschke, Glasgow
J McGill, Geneva
                    June 1976

CONTENTS

1   INTRODUCTION

Systems engineering, defined as the integration of available control
techniques to satisfy the requirements of a given control function, cannot
really be described as a discipline. The control engineering aspect (cyber-
netics) is probably the only part which has an underlying theory that can be
rigorously formulated and applied as such to practical problems. The digital
control of sequences of operations, in which the operations have a separate
existence, is to a large extent dependent on the nature of the operations
themselves. It can be given theoretical treatment only in very idealised
situations. Yet there is a practical need for a text which gives guidance
on such problems and at the same time is more than a catalogue of available
techniques.

As the available control hardware increases in power, becomes more
abundant and reduces in cost, there is a tendency to let it dictate the
control mechanism which will be used to perform a given function. That such
an approach is wrong is self-evident. Therefore an attempt to rationalise
the techniques of digital control is worthwhile even if it does not amount
to a comprehensive theory. What can be given is

1   a method for the rigorous formulation of a control task,
2   practical examples of such formulations, taken through to the point
where an appropriate control unit could be built,
3   a description of the relevant control techniques which explains both
their potential and limitations.

Thus the treatment must bring together subject matter which is normally found
separately in specialist literature. What has made this feasible, is the
availability of a wide range of control hardware in integrated modular form.
Thus it is now possible for complex technical means, such as electronic
devices, to be included in a design without first giving a complete introduction to the technology and science on which the devices depend. Fund-
amentals should never be ignored, but it is unnecessary to describe all the
internal details of a module before considering its incorporation in a
control system.

2

1.1   SCOPE OF THE TEXT

The purpose of the text is to describe a range of techniques which are
employed in industrial control operations.

A conceptual appreciation of the techniques is given,rather than a
series of recommendations on their applicability in given situations. This
is appropriate since the requirements are so completely dependent on the
process to be controlled and cannot be generalised.

For the same reason bounds have to be set on the part of the control
system which can be encompassed in the description. Fig 1.1 is a block
diagram of a complete system and shows which parts are covered in the text
and which parts are outside its scope.

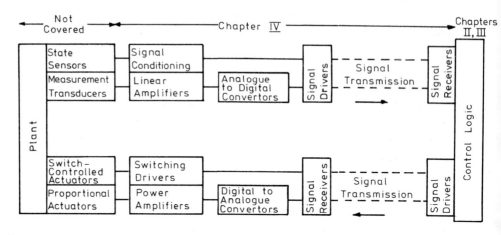

Fig 1.1   Complete control system showing coverage in the text.

There is a further restriction on the type of industrial process whose
control is envisaged. The techniques are aimed at the control of processes
which are made up of sequences of operations of essentially fixed duration.
There may be several such sequences which run contemporaneously and are
dependent on one another. The dependence is such that the outcome of one
operation determines the choices within another operation or the selection
of alternative operations from a sequence.

This contrasts with continuously variable processes in which the conditions under which a process is carried out are subject to adjustment by
controlling parameters. These parameters are functions of the results of
measurements carried out on the process. The control theory which is applic-

able here, in which the plant is represented by its transfer characteristic in a closed loop system is completely outside the scope of the present text.

Having defined the types of control to which the techniques apply a certain bias towards electronic techniques is justified. The control logic (Fig 1.1) is concerned with the logistics of the control. Apart from responding to external instructions, this means processing data and signals from the plant and determining therefrom the choice of subsequent actions. It may also have a data logging function and do a certain amount of computation. In all this, the techniques of electronic logic are by far the most convenient and the most powerful.

Outside the control logic the choice of technique is not so one sided. Electronic techniques have the disadvantage that signal communication must be protected from interferences due to the 'noisy' environment and from crosstalk between signal lines. While there is an obvious advantage in using electrical cables as lines of communication, the longer they are the more serious does the problem of interference become. The corruption of signals extracted from the plant will mislead the control logic, no matter how carefully it has been designed. The discovery of transmission errors is not sufficient in itself, since the control logic must have the necessary information to keep the plant functioning properly. On the actuator side it could be argued that it is best to convert logic signals into power signals and transmit the latter (at the cost of slightly heavier cables). However even this is generally unacceptable since it brings sources of electrical interference back into the sensitive area of the control logic.

The alternative of changing medium immediately outside the control logic must certainly be considered for slow processes. Hydraulic or pneumatic transmission, actuation and even sensing (if not data transmission) is a valid alternative. The inconvenience and extra cost of handling a more difficult medium could easily be justified by the increased reliability and robustness of the arrangement.

The omission of non-electronic techniques is not to be interpreted as meaning that they play a subsidiary role. The problem is that any treatment of hydraulic or pneumatic components would have to lay much greater emphasis on practical engineering consideration. There are less subtle difficulties but many more practical difficulties in selecting elements of the right size and capacity and finding suitable mechanical arrangements for their disposition and interconnection. The latter can only be assessed in a given situation and it is of doubtful value to describe the techniques in a general

sense. It is better to refer the potential user to the manufacturers' data
books, many of which give useful introductions to the technology.

What is not advocated is the use of hydraulic, pneumatic or even fluidic
devices in the control logic where the procedures are of necessity complex
and involve some form of data processing.

Chapter 2 defines one of the main objectives of the text. This is to
show how the requirements of the control operation can be translated into a
form in which it can be executed with the available hardware. The starting
point here is the control algorithm, which is orientated entirely towards
the application in hand. Since the latter is best understood by the plant
engineer rather than the specialist in control technology, the suggested
methods of formulating the algorithms should help to bridge a gap which is
often the stumbling block in the implementation of control systems.

1.2   APPROACH TO THE DESIGN OF CONTROL SYSTEMS

The wholesale automation of an industrial plant in one step is seldom
a practical proposition even in the case of a new plant. This leaves the
problem of planning piecemeal developments of the control system in such a
way as to avoid contradictory requirements arising in subsequent extensions.

This is one of the reasons why a decentralised control operation is
favoured. The cycle of operations to be performed in any unit of the plant
is embodied in an autonomous program controller which can be developed
independently.

The centralising function then reduces itself to

1   an interlock system which synchronises individual program controllers
at points of interdependence,

2   the organisation and the storage of data which is prepared in one part
of the control system and used in another,

3   the distribution of control signals from a central control console,

4   the collection of data for logging and signals for control panel display.

There is always a point in the planning of a control system when
decisions on the mode of implementation have to be made with little fore-
knowledge of the repercussions. It is also important to know at the outset
what technologies will have to be used. A list of considerations which
influence this is therefore given. They are:

a   The physical location of parts of the process to be controlled,

b   The speed with which events take place (cycle times),

c   The sensors and measurements to be used,

d   The mechanisms which must be operated and controlled,

e   The degree of interdependence of parts of the process,

f   The amount of data which has to be handled and stored,

g   Changes in operation which might prove necessary at a later stage,

h   Methods for continued operation in the case of failures,

i   The safeguards needed,

j   The method of installation, commissioning and fault diagnosis,

k   The environment in which the control system has to operate.

The following general comments can be made.

Problems of communication between different parts of the process multiply as distances inside the installation become large. Assuming that electrical communication is used, it will be seen (Chapter 4) that the techniques applicable within a range of say 1 meter are quite unusable over distances of tens of meters. Techniques for signal and data transmission which cope with longer distances are given (Section 4.2),but even these reach their limit. Carrier-based telecommunication techniques then become necessary. Communication problems must be considered at an early stage in the design of a control system.

The actual processes in a plant will generally be slow compared to the speeds at which the control logic can function. As an example,components might be handled in the process at a rate which is appreciably less than 100 per sec. Electronic speeds, even in the serial form advocated in the text, are 3 orders of magnitude faster. On the other hand some of the simpler interface elements such as relays and solenoids have a speed limitation not much higher than 100 operations/sec. Since they will be dedicated to individual actions, their speed would still be adequate in this case. Higher process speeds, however, change all this. The consideration of speed must therefore also figure early in the design.

The subject of sensors, transducers as well as operating mechanisms and mechanical drives (not covered in the text) is fundamental in any plant control operation. It is important to establish first whether the measurements at the required speed and to the required accuracy and resolution are physically possible. Similarly sensors, such as photocells observing passing components, must have a large enough signal to give reliable information.   If difficulties in this area cannot be resolved, it is pointless to proceed further.

Mechanisms, such as would be used to handle components, must be designed so as to malfunction as rarely as possible and must have a long life They should be able to cope with grossly faulty components without permanent damage.   The final performance of the control system will depend on the quality of the mechanisms used in the actual plant.

The next point to be considered is to what extent and in which way the control system can be decentralised.   The advantages of decentralisation are obvious.   Individual parts of the process may be developed and proved independently.   Any failures will be localised and manual intervention facilitated.   Fault diagnosis becomes much easier.   On the other hand,there is a limit to the complexity of the interlocks and to the amount of data interchange which it is practical to handle.   Hence the degree and nature of the decentralisation must be decided in any particular application.

There will be parts of the process in which the logistics consist of no more than a sequence of signals leading to action and vice versa.   An asynchronous method of control which has been used by Ashley and Pugh $\begin{bmatrix} 1.1 \end{bmatrix}$ will be referred to in Section 1.3.   The technique could prove to be the right one in certain parts of the plant.   It has much to commend it, particularly in slow processes, if the control can be implemented in low speed relay or fluidic logic.   The control would then be much less sensitive to disturbances.   An asynchronous system has no control over its response to signals from the sensors.   If however the logic elements only recognise genuine signals,this does not matter.

At the other extreme is a centralised on-line computer system (Section 1.4) to which all signals are routed and from which all actions are directed.   An on-line programmable computer which handles interrupts, receives data and performs all the data processing in its own order code can be an alternative to the special purpose hardware.   The main advantage is the consideration of subsequent program modifications,which are obviously much easier than the rebuilding of special purpose hardware.   However the penalties are considerable.   If a number of processes run simultaneously and their cycle times are comparable, the permutations of interrupts which can occur become numerous and very sophisticated routines are needed to react correctly to all situations.   The general purpose computer is therefore regarded as more appropriate to a data logging and storing function, which might involve the computation of statistics relevant to the process.   It would still be on-line, but function as a peripheral to the control logic.

It is wrong to dismiss the advantages of any form of control because of a preference for a homogeneous set of techniques for the whole control system. Therefore the following two sections will be devoted to a brief description of asynchronous and on-line computer control which do not figure in the main text. Full treatment of these subjects will be found elsewhere.

1.3   ASYNCHRONOUS CONTROL

description

A mechanism goes through a sequence of independent movements. The end of each movement signifies the start of the next.

Microswitches or other such devices are used at both ends of each movement. Electrical,pneumatic or hydraulic actuators provide the movements.

Such a mechanism can be driven simply by making each movement dependent on a logical combination of the end of movement signals. The only memory is the state of the signals themselves and there is no other form of timing. The change of state of a signal is called an event, the movement an action. Events result from actions and actions result from events. The mechanism, once started, will continue operating until some signal or some actuator is inhibited.

In the simple case, which will be considered here, movements are entirely sequential. Therefore only one signal can change state at any time. A control which depends on such signals is said to have monostrophic properties. It cannot deviate from the planned sequence of actions. Delay or contact bounce in the signal merely causes a delay or hesitation in the next action. There is no danger of transient combinations of signals which do not correspond to the planned state of the signals either immediately before or immediately after an event.

A simple example serves to illustrate the technique*. A pick and place mechanism starts with its transfer arm retracted. It lowers its grasp hand, closes the grasp, lifts the hand, extends the arm, lowers the hand, opens the grasps, lifts the hand, retracts the arm and repeats the cycle. The actions are defined as

A   reach down

B   grasp close

C   transfer extend

---

*It is the same example as used by Ashley and Pugh [1.1] . The symbols are kept the same for ease of comparison with this paper.

8

The actuators are energized in one sense when A, B, C are true and reversed (or de-energized in case of single-acting mechanisms) when A, B, C are false. For the latter the Boolean notation $\bar{A}$, $\bar{B}$, $\bar{C}$ is used to denote the negates of A, B, C. The absence of a third state implies that it requires a physical stop to the movements, so that an actuator can remain energized after the end of a movement. The corresponding limit switches are

P	reach mechanism up
Q	reach mechanism down
R	grasp mechanism open
S	grasp mechanism closed
X	transfer arm retracted
Y	transfer arm extended.

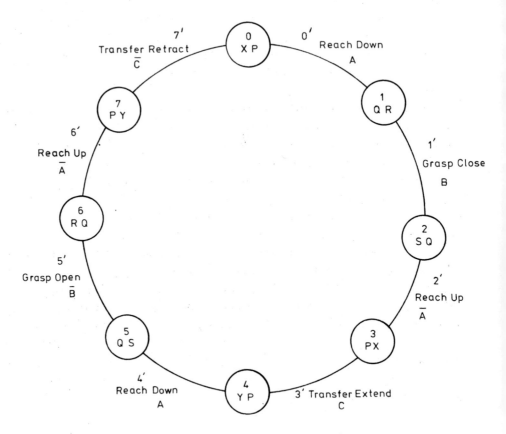

Fig 1.3.1 Cycle of operation. Pick and place mechanism.

The signals are true when a movement is at the limit and false (name barred) away from it.

Fig 1.3.1 shows the cycle of operation. The events are shown as circles numbered 0 to 7. The connecting actions are numbered accordingly 0' to 7', where 0' connects 0 and 1, 1' connects 1 and 2 and so on.

The occurrence of an event coincides with the closing of a limit switch. For example event 2 is due to the change $\overline{S}$ to S (grasp closing to grasp closed). The end of the event is a limit switch being broken. For example the end of event 2 is the change Q to $\overline{Q}$ (reach mechanism down to moving up). Therefore two letters are associated with every event as shown in Fig 1.3.1.

Associated with every action is the actuator which is energized when the action commences. The letter is written accordingly alongside the action, but it must be understood that the actuator remains energized even after the end of the action.

The corresponding sequence diagram is given in Fig 1.3.2. Here the limit switch signals P, Q, R, S, X, Y are shown against the sequence numbers 0 to 7 and the actuator signals A, B, C are given below them.

It remains to derive A, B, C from P, Q, R, S, X, Y using the Boolean operators ∧(AND) for conjunction, ∨(OR) for disjunction, as in 2.2.1.

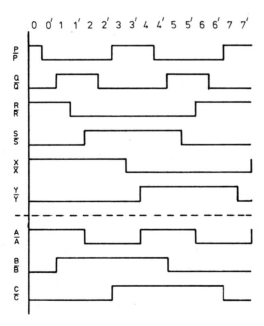

Fig. 1.3.2. Sequence diagram corresponding to cycle diagram (Fig. 1.3.1.)

The formulae required,can be distinguished almost by inspection in such a simple case as Fig 1.3.2. A few observations however help to make a systematic derivation clear.

(i)   Transitions of A, B, C, take place at the beginning of an event which means the transition of sequence number n' to n + 1.

(ii)   Transitions of P, Q, R, S, X, Y from the signal barred to the signal also take place at the beginning of an event

(iii) The reverse transition to ii (signal to signal barred) takes place at the end of an event which means the transition n to n'.

The simplest case is that of A, which follows immediately. A is formed by the signal whose rise is coincident with the rise of A in conjunction with the negate of the signal whose rise coincides with the fall of A.   Thus

$$A = (X \wedge \bar{S}) \vee (Y \wedge \bar{R}).$$

The case of B and C is slightly more involved. The rise of B coincides with the first transition $\bar{Q}$ - Q, the fall with the second transition. Hence the conjunction with a signal which changes state between the first and second transition is needed. $\bar{Y}$ is such a signal. Next there is the transition Q - $\bar{Q}$ while B remains true. Here the conjunction of $\bar{Q}$ with a signal which has changed just before the first transition Q - $\bar{Q}$ and is changed back before the second transition is needed. A signal which satisfies this condition is S.

Therefore B = $(Q \wedge \bar{Y}) \vee (\bar{Q} \wedge S)$

and by a similar argument C = $(P \wedge S) \vee (\bar{P} \wedge Y)$.

The control system requires no ancillary logic whatever. If 2-pole microswitches are used (1 NO + 1 NC contact), they need only be wired as in Fig 1.3.3 to generate the actuator signals $\bar{A}$, $\bar{B}$, $\bar{C}$*.

It may of course be impractical to have solenoid currents flowing directly through microswitch contacts. In that case relays are required, whose coils are in the microswitch circuit and whose contacts are in the solenoid circuit.

Whatever form the practical solution takes, there is no need to resort to high speed logic. Thus there will be no effective response to transient disturbances on the lines connecting the microswitches to the logic. More-over any spurious signals that carry sufficient energy to make a relay,are

---

*Connection to earth,instead of to supply,is preferred at the microswitches and actuators are energised by earthing signal lines.

unlikely to change the state of its contacts long enough for the solenoids
to respond.

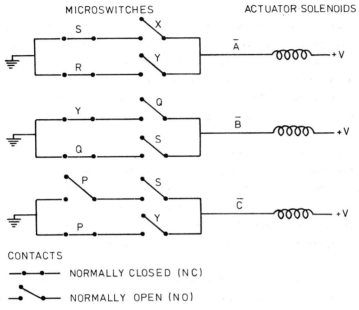

Fig 1.3.3 Control wiring for pick and place mechanism.

What has been described is a very satisfactory way of controlling a
process, as long as there is no choice of alternative sequences which depends
on data derived from the process. Beyond this point the techniques described
in the subsequent chapters have to be used. A number of new complications
are thereby introduced.

As an example, solid state memory elements, whose setting is controlled
by gating signals from the plant, are extremely vulnerable to spurious signals on the communication lines. Memory elements can make the most transient
disturbance look like a genuine signal. It is therefore good policy to gate
the signals with a synchronising pulse which comes up just prior to the time
when a valid signal could occur. Moreover such a pulse should be of the
shortest practical duration. In this way the acceptance of a spurious signal, though occasionally inevitable, is made less likely by reducing the
probability of the signal coinciding with the gating pulse.

One concludes that sophistication in control operations has to be fully
justified before accepting the consequences of the more powerful but also
more hazardous form of synchronous control.

1.4   ON-LINE COMPUTER SYSTEMS

Certain techniques - generally grouped under the term 'timesharing' - have been developed to allow a single computer* to work on a number of independent data-processing tasks at the same time,by giving attention to each in turn and overlapping computation with input-output (I/O) activity. In industrial automation, the problem of controlling a number of parallel operations simultaneously is similar to that found in scientific or commercial data-processing, but differs in the following important respects.

(i)   The computations performed are generally simple.
(ii)   The signal processing (I/O) section of the computer will be larger in comparison with that of other data processing systems since, even in a small automation project, the number and variety of signals involved is considerable.
(iii) Although many parallel operations may be involved in the process, the individual operations will often be quite simple and involve only a few program steps
(iv) Certain critical areas of control may require rapid response from the control computer. In conventional data processing applications,momentary loss of response may be tolerable and can almost always be recovered later. This is not true in a real-time control application involving mechanical motion, such as the positioning of a tool or workpiece.

The process to be analysed is first split up into its component control sequences. These will be coded as subprograms in the complete control program. The progress through each sequence will be determined by status signals from the plant and the results of measurements of plant conditions, as well as the status of related sequences. Up to this point both hardware and software solutions take the same path. Beyond this, however, there is the fundamental difference between the inherent parallelism of the hardware control units described in Chapters 2 and 3 and the serial form of control which has a common origin in the computer.

The simplest way of making the computer appear to exercise simultaneous control over each aspect of the process is to employ a supervisory program (scanning supervisor) which will execute each subprogram in turn in the

---

*The arguments developed in this section apply in general to both minicomputers and microprocessors. These are functionally identical. The microprocessor is the central processor (usually in Integrated Circuit form) of a minicomputer. It does not incorporate the interfaces which are normally associated with computers.

following way. As each subprogram is entered, a check is made to see if conditions are set to proceed to the next step in the control sequence. If they are set, instructions are executed until held up by another condition; if they are not set, rather than wait for status signals to change, the next subprogram in the cycle is entered. Communication and interlocks between programs use particular computer memory locations which hold binary indicators.

This approach has some drawbacks. It is slow and inefficient since a complete scan cycle may be required between successive steps in a sequence if an interrogation of status signals occurs between them. Also the sub-programs are entered irrespective of necessity. The supervisor is, however, easily programmed and is quite adequate where response times are not critical.

The alternative approach could be called 'event driven'. Here the change in state of signals from the process triggers the execution of the appropriate subprogram. This is done by means of the interrupt system of the computer which causes it to suspend and store the state of its current operation, identify the source of the interrupt and branch to the correspond-ing routine. This is obviously more complicated than the scanning program described above and requires that the computer be built with the following features.

(i)   Capacity for accepting a large number of interrupt signals.
(ii) Ability to inhibit (mask) each interrupt signal individually, so as to select only those interrupts which are pertinent to a subprogram at any instant.
(iii) Identification of interrupt source by hardware. Testing each interrupt source one at a time (flag polling), under program control, is inefficient for a large number of interrupts.
(iv) A multilevel interrupt structure to give priority to those events which require attention more urgently than others.

With these facilities no supervisory program is required to schedule the control sequences. However, routines must be provided to mask/unmask the appropriate interrupts as the control sequence proceeds and to resolve possible conflicts between subprograms.

The advantages of on-line computer control are clear. A well-designed program is easily modified if the process changes or expands. The computer is ideally suited to the calculation and data logging aspects of the problem.

It does however require special purpose software*. Particularly demanding
is the problem of providing adequate test and maintenance routines and of
ensuring continued operation of the plant when an actuator or sensor has
failed. The software required to cope with these conditions may be very
complex. A distributed control,using the techniques which will be described
later, makes it possible to incorporate these features more easily.

---

*While high level languages could be used in some parts of the program, much
of it must be written in machine-dependent assembly language, since so many
functions are tied to the structure of the input-output mechanism.

2   FORMULATION OF ALGORITHMS

The starting point for the representation of a control algorithm is normally a flowchart or block diagram. It gives a bird's eye view of the process by condensing minor detail into appropriately described blocks. Such a representation is generally useful, but the question arises whether it is in an adequate form for the next step in the design procedure.

If the implementation is centred on a programmable general purpose computer there are good reasons for the block diagram presentation.

One reason is the centralized nature of the processing. The block represents a sequence of instructions to which the computing unit is dedicated, together with all working store, whenever it is called. As soon as a number of computing units operate simultaneously the block diagram form is unenlightening.

Another reason for the use of block diagrams is the construction of a program from subroutines. Some of these may be incorporated in library systems and hence need not be described in detail. Others will be detailed separately with their own flowcharts so that no one diagram becomes excessively complicated.

If the object is the construction of a special-purpose computer/controller the above considerations are no longer relevant or important.

Single computing units are seldom used, since it is generally more convenient to have a multiplicity of independent computing units, each of which is called into action when required and allowed to function in parallel with others.

The use of subroutines is not entirely irrelevant. It will be shown in Section 2.6 that the use of closed subroutines is easily realizable in hardware and is sometimes very powerful. However they are not used as extensively as in a computer program (again largely because of parallel processing) and there are no standard subroutines in hardware for inclusion in a controller.

The main reason however for advocating a different method of formulating the algorithm is the need for complete rigour in the representation, since there is no other (coded) form to arbitrate on points of detail.

The approach which will be adopted is similar to that advocated by Iverson [2.1] in 1962. In his book 'A programming language' (APL) Iverson formulates a symbolic language for the description of algorithms which is

both concise and definitive. Much of its rigour will be dispensed with, but
the general principles are well suited to the translation of an algorithm
into special purpose hardware. To support the use of such symbolism the
following paragraph from the introduction to the book is quoted.

'Ordinary English lacks both precision and conciseness. The widely used
Goldstine-von Neumann (1947) flowcharting provides the conciseness necessary
to an overall view of the process only at the cost of suppressing essential
detail. The so-called pseudo-English used as a basis for certain automatic
programming systems suffers from the same defect. Moreover the potential
mnemonic advantage in substituting familiar English words and phrases for
less familiar but more compact mathematical symbols fails to materialise
because of the obvious but unwanted precision required in their use.'

2.1 DEFINITIONS

Boolean vectors and matrices will be used to represent the bit configuration of a store and set of stores respectively. The notation conven-
tions of matrix algebra will be applied but with the following variations.
(i)   A vector $\underset{\sim}{a}$ will be considered a row vector to correspond to the general
convention of drawing stores horizontally. Where the algebra requires a
column vector $\underset{\sim}{a}'$ will be used to indicate transposition.
(ii)   The elements of a matrix A, whose row suffix is i and column suffix j
will be denoted by a^i_j. The corresponding elements of the transposed matrix
A' are a^j_i.
(iii) Suffices will run from the highest (one less than the number of components) to the lowest (0) from left to right, contrary to the normal
convention.

The reason for the last convention will be obvious from an example.
If the register B, of dimension τ contains a vector $\underset{\sim}{b}$, the value of its
content is obtained by scalar multiplication with a weight vector $\underset{\sim}{s}$ whose
components are $(2^{\tau-1}, 2^{\tau-2}, \ldots, 2^0)$,

$$b = \underset{\sim}{b}\underset{\sim}{s}' = b_{\tau-1}2^{\tau-1} + b_{\tau-2}2^{\tau-2} + \ldots + b_0.$$

The vector $\underset{\sim}{b}$ 'represents' the number b. The components of the vector
are written with the most significant $(b_{\tau-1})$ first, the least significant
(b_0) last, as in ordinary number representation.

The contents of registers are the operands of the algorithm. Operations
on the registers are carried out by a number of basic operators which are
incorporated in the circuitry. More complex operations are called derived

operations and take the form of subprograms. Some of these will be described in Section 2.3.

The basic operations include the Boolean operations, addition, sub-traction, counting, scaling and register transfers. Since the execution of the operations is to be controlled by a synchronising clock, there is a distinction between two modes of execution: parallel and serial.

In parallel operation vectors are operated upon by providing a gating operator for each component of the vector, so that the operation is completed in one clock period. For example the operation which transfers $\underset{\sim}{b}$ to $\underset{\sim}{c}$, denoted by $\underset{\sim}{c} \leftarrow \underset{\sim}{b}$ consists of replacing every component of $\underset{\sim}{c}$ by the correspond-ing component of $\underset{\sim}{b}$.

The slightly more complicated operation,

$$\underset{\sim}{c} = \underset{\sim}{c} + \underset{\sim}{b},$$

in which the addition of two components may produce a carry into the next more significant component is represented diagrammatically in Fig 2.1.1.

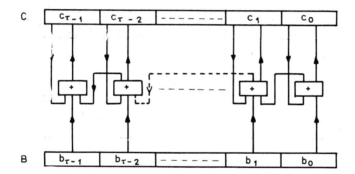

Fig 2.1.1 Parallel addition.

The multiplicity of gating operators is only justified if the speed of operation demands the completion of the operation in one clock period or if the number of components is small.

The alternative is serial operation in which each component is passed through a single gating operator sequentially in time. If the dimension of the vector is τ, the operation is completed at the end of τ clock periods. The time lapse of τ clock periods is called the 'Characteristic Period' of the serial operation, henceforth to be known simply as the Period.

The basic storage element for serial operation is the shift register. The vector $(\underset{\sim}{c})$ is here continuously right circularly shifted. The weight of

18

component c_k depends on the value of a synchronizing count $\underset{\sim}{t}$ which is counted so that successively $t = 0, 1 \ldots \tau - 1, 0, 1 \ldots$ and so on. At $t = 0$, c_k has weight s_k, at $t = 1$ weight s_{k+1} and in general the weight of c_k is s_{k+t}. Since the shift is circular, a modulo notation is needed to allow for the repetition of the same weighting after τ clock periods. Using the Iverson notation $m|n$ for n modulo $(m)^*$, the weight of c_k in a shift register of length τ is $s_{\tau|k+t}$.

Fig 2.1.2 shows the content of such a shift register at three different times during the Period.

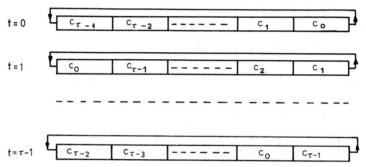

Fig 2.1.2 Shift register content.

The gating operator is inserted into the circulation between the least significant and most significant stage of the register. In fact these are generally the only access points to the register. The simplest case is the transfer $\underset{\sim}{b}$ to $\underset{\sim}{c}$. This means making the input to the most significant stage of C the output of the least significant stage of B instead of recirculating $\underset{\sim}{c}$.

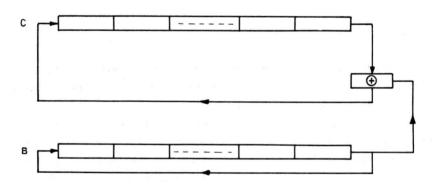

Fig 2.1.3 Serial addition of operands.

*Residue 0, see Iverson (1961) $\left[2.1\right]$.

Fig 2.1.3 shows the serial execution of an operation involving 2 operands such as the addition $\underset{\sim}{c} = \underset{\sim}{c} + \underset{\sim}{b}$. The gating arrangement labelled $\oplus$ is slightly different from that used in Fig 2.1.1 since it includes the storage of the carry from one clock period to the next.

To be able to use the vector notation, a selection vector has to be defined as follows. A unit vector $\underset{\sim}{\varepsilon}^k(\tau)$ has τ components in which the component with suffix k is a 1 and the remainder 0. If the value of k depends on a time parameter t, a family of unit vectors may be defined by the selection vector $\underset{\sim t}{\varepsilon}$, where $\underset{\sim t}{\varepsilon} = \underset{\sim}{\varepsilon}^k$ when t = k. The vectors are:

$$\text{for } t = 0, \varepsilon^0 = \ldots\ldots\ldots 0001,$$

$$t = 1, \varepsilon^1 = \ldots\ldots\ldots 0010,$$

$$t = 2, \varepsilon^2 = \ldots\ldots\ldots 0100 \text{ etc.}$$

The transfer of $\underset{\sim}{b}$ into $\underset{\sim}{c}$ can now be written formally as

$$\underset{\sim}{c} = \underset{\sim}{c} \vee (\underset{\sim}{b} \wedge \underset{\sim t}{\varepsilon})$$

where $\vee$ and $\wedge$ denote the logical operation OR and AND respectively and $\underset{\sim}{c}$ is assumed to have its previous content cleared. The $\wedge$ operation selects the components of $\underset{\sim}{b}$. The $\vee$ operation signifies that wherever and whenever the component is 1, it is inserted into $\underset{\sim}{c}$, otherwise $\underset{\sim}{c}$ is unchanged.

The formulation does not specify whether $\underset{\sim}{b}$ is held in a parallel register or shift register. In the former case gating with ε_t serializes the components. In the latter case they are already available serially from the least significant stage of the shift register.

Similarly C may be a parallel or serial register. In the former case $\underset{\sim}{c}$ is reset initially and then has the components of $\underset{\sim}{b}$ inserted as a parallel operation. In the latter case the input to the most significant stage of C receives the components of $\underset{\sim}{b}$, while the previous content is removed.

With this understanding the notation $\underset{\sim}{c} \leftarrow \underset{\sim}{b}$ will be used to denote the transfer and $\underset{\sim}{c} \leftrightarrow \underset{\sim}{b}$ an interchange of contents.

Similarly for any other operation (op), $\underset{\sim}{c} \leftarrow \underset{\sim}{c}$ op. $\underset{\sim}{b}$ will be used for

$$\underset{\sim}{c} \leftarrow \underset{\sim}{c} \text{ op. } (\underset{\sim}{b} \wedge \underset{\sim t}{\varepsilon})$$

One extension to the notation will be used. If, for example, the selection vector $\underset{\sim t}{\varepsilon}$ in the above example is replaced by $\underset{\sim t-1}{\varepsilon}$, all components are selected from $\underset{\sim}{b}$ one clock period later. The result is that their weights are doubled.

Thus c = 2b is obtained from

$$\underset{\sim}{c} \leftarrow \underset{\sim}{c} \vee (\underset{\sim}{b} \wedge \underset{\sim t-1}{\varepsilon}).$$

The notation $\underset{\sim}{c} \leftarrow \underset{\sim}{b}_{-1}$ will be used to show such scaling. In general $\underset{\sim}{c} \leftarrow \underset{\sim}{b}_{\pm r}$ results in $c = 2^{\mp r}b$.

Some operations are realised by using different synchronizing counts, so that not all the registers are synchronized to the same (t) count. In such cases the shorthand notation is inadequate and it will be necessary to revert to the more rigorous definition with explicit selection vectors.

It remains to show how the parallel storage of a vector is converted into the serial form and vice versa. The first is referred to as multiplexing, the second as demultiplexing.

If, as an example, an 8-component vector is involved, a timing count $\underset{\sim}{t}$ of dimension 3 is required for the conversions. The sequence of values taken by this count is 0 1 2 3 4 5 6 7 0 1 2 ... .

For the parallel to serial conversion a <u>multiplexer</u> has its parallel inputs connected to the components of the vector in the parallel store. The multiplexing inputs are taken from $\underset{\sim}{t}$ and the serial ouput feeds the circulation of the serial store. The arrangement is shown in Fig 2.1.4. The shift register is loaded in one Period (t_0 to t_7 inclusive) during which LOAD is true, while the normal circulation is blocked by $\overline{\text{LOAD}}$.

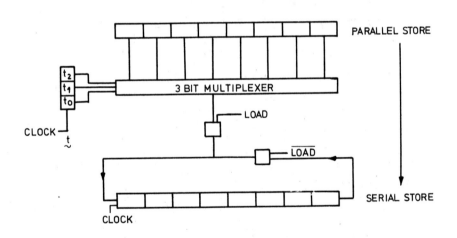

Fig 2.1.4 Parallel to serial conversion.

The reverse process of serial to parallel conversion requires a <u>demulti</u>-<u>plexer</u> or <u>decoder</u>. The encoded input is taken from $\underset{\sim}{t}$ and the decoded output goes to the parallel store. The shift register is connected by its least

significant stage to the decoder ENABLE which gates all the decoder outputs.
The parallel store is initially reset. If the enabled decoder output becomes
1, the stage of the parallel store connected to it, is set. If the output
is 0, the stage remains in its previous state*.

A Period is needed for the conversion. Fig 2.1.5 shows the arrangement.

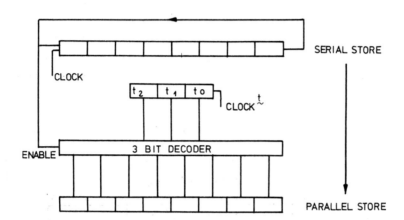

Fig 2.1.5 Serial to parallel conversion.

2.2 BASIC OPERATIONS

Operations are broadly classed as Boolean or arithmetic; subsections
2.2.1 and 2.2.2 deal with Boolean and arithmetic operators separately. In
the subsection 2.2.3 (conditional arithmetic operations) Boolean and arithmetic operations will be combined with the following understanding. If a
Boolean operand remains unchanged throughout an entire arithmetic operation,
it can be used to perform a Boolean operation on the latter.

For example, if a and b are two arithmetic operands, their addition
when α (Boolean) is true and subtraction when $\bar{\alpha}$ is true can be written as
$[\alpha \wedge a + b] \vee [\bar{\alpha} \wedge a - b]$. This means simply that the adder/subtractor is
set to one or the other state depending on α.

2.2.1 Boolean and relational operations

Boolean operations are concerned with Boolean operands which can only
take one of two states, true or false.

---

*Where this mode of setting is not available on a parallel store a decoding
(addressable) latch can be used in place of the simple decoder.

Relational operations perform comparisons on arithmetic operands and give a result which is Boolean. The latter becomes a Boolean operand for subsequent operations.

A Boolean operand is held in a single bistable. Boolean arrays are held in parallel bistables whose inputs and outputs are accessible independently. The use of shift registers is not favoured in this context.

Boolean operations

If α and β are two Boolean operands, $((2)^2)^2$: 16 distinct Boolean operations can be performed with them. Table 2.2.1 gives the complete list and the symbolism which will be used*.

Table 2.2.1 Symbolism for Boolean operations on 2 operands α and β.

α β	OPERATION NUMBER
	0 1 2 3 4 5 6 7 8 9 10 11 12 13 14 15
0 0	0 1 0 1 0 1 0 1 0 1 0 1 0 1 0 1
0 1	0 0 1 1 0 0 1 1 0 0 1 1 0 0 1 1
1 0	0 0 0 0 1 1 1 1 0 0 0 0 1 1 1 1
1 1	0 0 0 0 0 0 0 0 1 1 1 1 1 1 1 1

Operation number	Symbol	Name
0	redundant	
1	⊽	NOR
2	↚	negated reverse implication
3	redundant	
4	↛	negated implication
5	redundant	
6	≢	not equivalent
7	⊼	NAND
8	∧	AND
9	≡	equivalent
10	redundant	
11	⟶	implication
12	redundant	
13	⟵	reverse implication
14	∨	OR
15	redundant	

*An ambiguity in the use of ⟵ (reverse implication) is tolerated since it occurs rarely and is unlikely to be confused with the symbol for transfer.

It is well known that the operations given in Table 2.2.1 are far from independent. They can all be reduced to one operation (say ∧ or ∨) operating on α , β and their negations ᾱ, β̄ . In fact the Sheffer Stroke-notation even reduces these two operations to one. However the understanding of the operations is often made easier by using the separate symbols given in Table 2.2.1.

The only operation which may need some explanation is the implication denoted by →. This suggests that α false has always a true implication, namely either β false or true is implied. How this can be was the question addressed by the Scottish philosopher McTaggart to the Cambridge mathematician G.H.Hardy. He asked 'if twice two is five, how can you prove that I am the Pope'? The answer given was 'If twice two is five, then four equals five. Subtract three from each side then one equals two. But McTaggart and the Pope are two. Therefore McTaggart and the Pope are one.'

In other words if α is false nothing can be said about β. This characteristic of an implication will come to the fore again when conditional out of sequence entries are dealt within section 2.4.

Relational operations

The six relational operations which define comparisons between two arithmetic operands are:

$$< \quad > \quad = \quad \not< \quad \not> \quad \neq$$

They are to be reduced to Boolean operations on the components of the arithmetic operands. The results of the comparison is formed in the Boolean β.

Fig 2.2.1 shows two arithmetic operands a and b held in shift registers A and B.

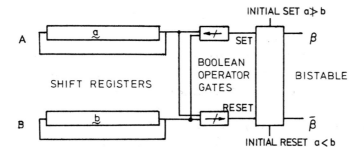

Fig 2.2.1 Serial comparison.

The least significant components of $\underset{\sim}{a}$ and $\underset{\sim}{b}$ are inputs to two Boolean operator gates. The one shown in line with register A performs the operation 2 of Table 2.2.1. The one shown in line with B performs operation 4. A **1** at the output of the gates is used respectively to set or reset the bistable which holds β. If β is given an initial setting its state at the end of the circulation gives the results of the comparison. The cases of a < b and a ≯ b (a ≤ b) are illustrated in Fig 2.2.1.

The complete set of relational operators and their Boolean equivalents are given in Table 2.2.2.

Table 2.2.2 Relational operators.

Relational operation	Set β if:	Reset β if:	Initial β
a < b a ≯ b	a ⊬ b	a ⊬ b	0 1
a > b a ≮ b	a ⊬ b	a ⊬ b	0 1
a = b		a ≢ b	1
a ≠ b	a ≢ b		0

The initial setting of β is needed since the Boolean operators give a setting of β for only two out of the four combinations $\overline{\overline{ab}}$, $\overline{ab}$, $\overline{a}b$, ab.

2.2.2 Arithmetic operations

Of the four arithmetic operations
+ addition   - subtraction   x multiplication   / division
the first two are basic, and will now be described, together with some related operations. Multiplication and division are derived from them and will be given in the form of subprograms in Section 2.3.

Before describing the operations the question of the representation of negative numbers has to be answered.

In a general purpose computer the signs of quantities which will be handled in the registers cannot be predicted. Hence allowance has to be made universally for the occurrence of negative quantities. The multiplier and divider must be programmed so that they can handle quantities of both signs.

The 2's complement representation is the most common. This means that a store of length $\tau + 1$ is needed to hold quantities in the range $-2^\tau \leqslant x < 2^\tau$. The most significant bit (2^τ) is then read as a sign bit which is 1 for negative quantities and 0 for positive quantities*.

In the case of special purpose computing and control equipment, the situation is different. The occurrence of negative quantities can be anticipated and in general will be much less common. Moreover at some stage control action will be required. As an example, a movement in positive or negative direction, according to the sign of a quantity, requires separately the numerical distance to be moved and a two-state indicator for the direction of motion.

For these reasons the alternative form of sign and modulus will be used for the representation of numbers.

For subtraction, where the relative magnitude of the operands is uncertain, the arithmetic operation will be preceded by comparison and, if necessary, reversal of operands. With this exception the choice of number representation simplifies the logic. All operands in the subsequent logic will be assumed positive.

Addition/Subtraction

In the following, addition and subtraction are expressed as Boolean operations on the components of the arithmetic operands. (As before, the shorthand $\underset{\sim}{c}_{-r}$ will be used for $\underset{\sim}{c} \wedge \underset{\sim}{\varepsilon}_{t-r}$)

$$\underset{\sim}{a} \pm \underset{\sim}{b} = \underset{\sim}{a} \neq \underset{\sim}{b} \neq \underset{\sim}{c}_{-1}$$

where $\underset{\sim}{c}$ (initially 0) is the carry derived from

$$\underset{\sim}{c} = \underset{\sim}{a} \wedge \underset{\sim}{b} \vee [(\underset{\sim}{a} \neq \underset{\sim}{b}) \wedge \underset{\sim}{c}_{-1}] \quad \text{for } a + b$$

$$\underset{\sim}{c} = \underset{\sim}{\bar{a}} \wedge \underset{\sim}{b} \vee [(\underset{\sim}{\bar{a}} \neq \underset{\sim}{b}) \wedge \underset{\sim}{c}_{-1}] \quad \text{for } a - b.$$

In serial operation the vector $\underset{\sim}{c}$ need not exist as such. It needs only a single bit storage which at time t is

set by $\qquad\qquad\quad \underset{\sim}{a} \wedge \underset{\sim}{b} \quad$ for addition

$\qquad\qquad\qquad\qquad \underset{\sim}{\bar{a}} \wedge \underset{\sim}{b} \quad$ for subtraction

and reset by $\qquad\quad\;\; \underset{\sim}{\bar{a}} \wedge \underset{\sim}{\bar{b}} \quad$ for addition

$\qquad\qquad\qquad\qquad \underset{\sim}{a} \wedge \underset{\sim}{\bar{b}} \quad$ for subtraction

and remains unchanged otherwise. It is used in the evaluation of $\underset{\sim}{a} \neq \underset{\sim}{b} \neq \underset{\sim}{c}$ at time t + 1 and then adjusted to its new state.

---

*In a few computers (for instance Univac) a 1's complement representation is used, which leads to a troublesome ambiguity in the centre of the range, viz the value 0 which has a different representation a +0 and -0.

Counting

Counting is only a special case of addition and subtraction in which the operand is 1 (vector ε^o). However it is more usual to use counters which may be incremented or decremented at the end of every clock period.

Counters are unsuitable for other arithmetic operations and it may be necessary to transfer operands between counters and registers or vice versa during the course of an algorithm.

Generally counters only control arithmetic operations and are not involved in them as operands.

This controlling function depends on the value of the counter, which will be denoted by a bracketed suffix. Thus $t_{(u)}$ is the Boolean which becomes true when $t = u$. Important in this respect is the termination of a modulo counter. A modulo (τ) counter returns to $t_{(o)}$ after $t_{(\tau-1)}$. $\tau - 1$ clock periods after $t_{(0)}$ the terminator $t_{(\tau-1)}$ is set. The terminator is reset at the end of τ clock periods.

A modulo (τ) counter is needed to synchronize serial registers of dimension τ and control program sequences which use them. The terminator $t_{(\tau-1)}$ plays a central part in the program sequence.

Scaling and floating point form

The operations of scaling, (multiplication by a power of 2) has been already explained.

One method of scaling is as follows. If the registers shift at a frequency ν, the controlling clock is derived from a clock of frequency 2ν which is divided to provide a two-phase (non-overlapping) clock (see Fig 3.1.8). Let the shift take place at the end of the second clock (clock odd). The first clock (clock even) is available for extra shifts. For every such additional shift the content of the register is right shifted, that is halved. Doubling or left shifting is obtained by an equivalent suppression of shifts.

Scaling is the key to the floating point form of holding variables. This form commonly used in computer arithmetic, but occurs less frequently in control systems.

In the floating point form, all variables are standardized such that a variable (say a) is held in two stores as the mantissa p and the exponent q, where

$$a = p2^q \quad \text{and} \quad \tfrac{1}{2} \leqslant p < 1.$$

After an arithmetic operation the result must be restandardized. A convenient way is to use one circulation to count the number of leading 0s

in the mantissa. In the next circulation this count is subtracted from q
and an equal number of shifts is suppressed in recirculating p.

Multiplication and division present no further problems. In addition
and subtraction however exponents must be aligned so that the addition

$$a + b = p2^q + r2^s$$

involves finding the smaller of q and s and recirculating its companion
mantissa with |q-s| additional shifts. The exponent of the result is the
larger of q and s and restandardizing may be necessary in the case of
subtraction.

It is also possible, in the case of addition, that the mantissa of the
result could become equal to 1. All registers must thus have an extra stage
at the most significant end to allow this to happen. An additional shift
and an increase of the exponent by 1 will return the result to the standardized form.

The obvious advantage of the floating point form is the determination
of register capacity by the number of significant figures in a variable
rather than by its magnitude.

The question of accuracy of floating point arithmetic is dealt with
in the many modern textbooks on numerical analysis $\left[2.2\right]$. The only general
comment worth making is that the presence of a certain number of significant
figures in the mantissa does not necessarily mean that all these are meaningful. Thus in a subtraction of two nearly equal quantities the result in
unstandardized form will have fewer **significant figures**. Restandardizing
does not change this.

Modulo

The notation $c|b$, b modulo (c), is used to denote the remainder of
$b \div c$. If $b \gg c$, division is the most appropriate way of obtaining
the remainder. If b and c are of the same order of magnitude, it is obtained
by successive subtractions of c from b and simultaneous comparison of the
difference with c until it is smaller than c.

A particular case occurs when $c = 2^r$. Assuming vectors of τ components

$$c|\underset{\sim}{b} \Leftrightarrow \underset{\sim}{b} \wedge \underset{\sim}{\bar{\alpha}}^{\tau-r}(\tau)$$

where $\underset{\sim}{\alpha}^s(\tau)$ is the masking vector of τ components whose s most significant
components are 1's and the remainder 0. Its negate is $\underset{\sim}{\bar{\alpha}}^s$.

As an example, let $\tau = 5$, $r = 3$, $\underset{\sim}{b} = 10101$,

then $\underset{\sim}{c} = 2^3 = 01000$, $\underset{\sim}{\alpha}^2 = 11000$, $\underset{\sim}{\bar{\alpha}}^2 = 00111$

$c|\underset{\sim}{b} = 10101 \wedge 00111 = 00101$, or numerically $8|21 = 5$

2.2.3 Conditional arithmetic operations

A stored program computer relies on a sequence of program steps for the execution of an algorithm. Every conditional alternative takes the form of a condition-dependent branch into alternative sequences. In a special purpose computer the serial execution of program steps is only needed where a particular program step depends on an earlier one in the sequence. Otherwise the algorithm can be implemented in parallel, as part of the circuitry, in the form of logical gating between the various registers and bistables.

It follows that many conditional operations do not require alternative sequences, but merely suitable gating logic, such as alternative inputs to an OR gate. The algorithm may set up a condition (say α) in one program step and follow this by an operation, such as for example,

$$d \leftarrow [(a + b) \wedge \alpha] \vee [(a + c) \wedge \bar{\alpha}]$$

where $\qquad \alpha \rightarrow (d \leftarrow a + b) \quad$ and $\quad \bar{\alpha} \rightarrow (d \leftarrow a + c)$

or $$d \leftarrow a + (-1)^{\alpha} b$$

where $\qquad \alpha \rightarrow (d \leftarrow a - b) \quad$ and $\quad \bar{\alpha} \rightarrow (d \leftarrow a + b).$

By a suitable formulation of the algorithm the number of program steps and alternative sequences can be kept to a minimum so that complex program controllers are seldom needed.

2.2.4 Array operands

A structured list of operands is called an array of operands. Since each operand is represented by a row vector, the elements $a_j^i, i=\mu-1,\ldots0;$ $j=\nu-1\ldots0$ represent an array of μ operands, each of which has ν components.

To address the operands the suffix i becomes the selection input to a multiplexer whose parallel inputs are the μ operands. The case of $\mu = 4$ is illustrated in Fig 2.2.2.

The effect of the multiplexer is to decode the vector $\underset{\sim}{i}$ so that the value of i determines which of the operands appears at the serial output.

If $\underset{\sim}{i}$ is held in a modulo (μ) count and the shift registers are synchronised to a j (modulo (ν)) count, operations like

$$\underset{\sim}{c}^i \leftarrow \underset{\sim}{a}^i + \underset{\sim}{b}^i$$

can be performed in one program step. The termination of the j count clocks the i count. The termination of the i count indicates the completion of the operation.

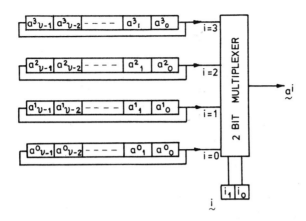

Fig 2.2.2 Addressing of serial array operands.

The use of shift registers means that only a single multiplexer is needed to address a two-dimensional bit configuration. One serialization is implicit and the multiplexer is used to address the one-dimensional array of vector operands.

If the two-dimensional array is held in static parallel form, such as in a core store, a two-dimensional multiplexer tree is required to access it. Fig 2.2.3 shows the case of a 4 x 8 matrix A, in which the rows are drawn adjacent to one another horizontally. The elements a^i_j are connected to a multiplexer tree. If the lower suffix (j) is varied more rapidly than the upper suffix (i), the result is to serialize the matrix into 4 operands of 8 components. A general form of store selection is shown diagrammatically in Fig 2.2.4.

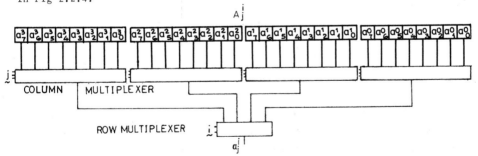

Fig 2.2.3 Addressing and serializing of parallel array operands.

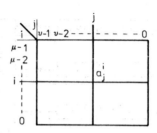

Fig 2.2.4 General representation of store addressing.

The element addressed is at the intersection of the lines which run through the decoded i and j values. A $\mu \times \nu$ matrix requires $\mu + \nu$ selection lines. It follows that the most economical arrangement for an n element store is a square matrix, when the number of selection lines is $2\sqrt{n}$.

2.3 DERIVED OPERATIONS

The method of deriving more complex operations from the basic operations of the last section will be illustrated with four examples.

The first two are the operators

x multiplication

/ division.

They are of particular value in demonstrating the formulation of an algorithm. However, where control systems require such arithmetic operations to any extent, the use of a microprocessor with a stored program is often more convenient [2.3]. Here a computer is stripped down to its essentials in order to execute what are very simple computer tasks. It is an alternative to building up a control system to perform in the computing area.

The other two examples are the code conversions

BDC - binary

binary - BCD.

They occur, at least in the integer form, at the interface between data input and output devices and the control system.

Derived operations are executed in a number of program steps which are brought together to form a subprogram. This may require program steps to be repeated cyclically, until some condition is satisfied which causes the program to exit from the loop.

The formation of a subprogram is illustrated below and explained in the next paragraph.

		ISE	OSE
P_0	Basic operations	$P_1 \wedge \delta$	
P_1	Basic operations	γ	
P_2	Exit	$\bar{\delta}$	$P_0 \wedge \bar{\gamma}$

Three program steps P_0, P_1, P_2 are assumed to be needed, of which P_0 and P_1 are cycled while δ and γ are true. If either δ or γ become false the subprogram proceeds to P_2 from which it returns to the main program.

The abbreviations ISE and OSE are used for In Sequence Entry and Out of Sequence Entry respectively. The entries are given as the conjunction of the entry condition and the program step from which the entry is made. Arrows on the left show backward OSEs, arrows on the right forward OSEs.

The subject of program structure has been anticipated here. It will be dealt with more fully in 2.4 and 2.6.

2.3.1 Multiplication

The product to be formed is

$$d = b \times c.$$

Arbitrarily c is taken as the multiplicand, b as the multiplier. If both b and c have dimension τ, the product d will have dimension 2τ. It will be formed in τ successive steps during which the components of b are examined in turn. If the component is a 1, c is added to d scaled by the appropriate power of 2.

The algorithm may be formulated in several alternative ways. The first choice is whether the components of b will be examined from the least signi-ficant up or most significant down. If the latter is chosen, the algorithm requires that at the rth step $d \leftarrow d + c_{r-\tau}$, if $b_{\tau-r} = 1$.

The successive scaling of b and d relative to c can be obtained by lengthening the circulations B and D by one component. A better alternative is to operate with two selection vectors ε_t and ε_q where q is derived from a count which is 1 shorter than the t count. Thus

$$\varepsilon_q = \varepsilon_{t+r}$$

where $r = 1, 2 \ldots \tau$ are the successive steps in the multiplication.

Since it is necessary to distinguish the most significant from the least significant half in the 2τ dimensional product d, the q count has to be a modulo 2τ count. The t count can remain a modulo τ count, since it is

only used for synchronisation. However, if it is assumed that t also runs
from 0 to $2\tau - 1$, the relation $q = t + r$ is obtained by double counting q
when $q = 0$.

Thus $\underset{\sim}{b}$ and $\underset{\sim}{d}$ are synchronized to the t count and are selected by the
q count. On the other hand the multiplicand $\underset{\sim}{c}$ must be synchronized to the
q count so as to remain unscaled. This is achieved simply by providing an
extra shift at the same time as the double count occurs. $q_{(2\tau - 1)} = t_{(\tau - 1)}$
transfers to the exit step and all stores are left synchronized to t (including the q count).

The program is given in Table 2.3.1, where it is assumed that upon
entry $\underset{\sim}{d}$ is clear and that $\underset{\sim}{b}$ and $\underset{\sim}{c}$ are set to contain respectively the multiplier and multiplicand.

Table 2.3.1 Multiplication $d = b \times c$.

$$P_0 \quad (\underset{\sim}{b} \wedge \underset{\sim}{\varepsilon}_t)_{q(\tau)} \rightarrow \left[\underset{\sim}{d} \leftarrow \underset{\sim}{d} \wedge \underset{\sim}{\varepsilon}_t + \underset{\sim}{c} \wedge \underset{\sim}{\varepsilon}_\tau \Big| q\right] q_{(\tau)}, q_{(\tau+1)} \cdots q_{(2\tau-1)}$$

$$\quad\quad (q \leftarrow q + 1)_{q(o)}$$

$$P_1 \quad \text{EXIT} \quad \dots\dots\dots\dots\dots\dots\dots\dots\dots\dots\dots\dots\dots\dots\dots q_{(2\tau-1)}$$

The following notes are added.

(i) $t = q - r$, when t is assumed to count from 0 to $2\tau - 1$.
(ii) During the first cycle, $r = 1$, the component $\tau - 1$ is selected from
the multiplier $\underset{\sim}{b}$ and components $\tau - 1, \tau$, ..., $2\tau - 2$ of $\underset{\sim}{d}$ receive components
$0, 1 \dots \tau - 1$ of $\underset{\sim}{c}$ and so on for subsequent cycles.
(iii) While the addend to the $\underset{\sim}{d}$ adder is admitted only from $q_{(\tau)}$ to $q_{(2\tau - 1)}$,
the addition continues for the full cycle since a carry may be propagated.

By tabulating the τ^2 steps of the operation in an actual example, the
formulation of the algorithm can be confirmed. The design of circuitry for
the operation is dealt with in Chapter 3 and the circuit for integer multiplication is one of the examples of 3.4.

So far only integral operands have been considered. Fractional operands
are needed when there are angular values (fractions of π) or trigonometric
functions. Mixed operands are much rarer in control algorithms and can
generally be avoided by scaling or by using floating point arithmetic (see
2.2.2).

The extension to fractional operands follows immediately provided negative suffices are admitted for fractional components.

If the suffices $0, -1, \ldots, -\tau + 1$ are used for components of weight $2^0, 2^{-1}, \ldots\ldots, 2^{-\tau+1}$, a new selection vector δ_t, where $\underset{\sim}{\delta}^k(\tau) = \underset{\sim}{\varepsilon}^{k-\tau+1}(\tau)$ replaces $\underset{\sim}{\varepsilon}_t$. At $t_{(o)}$, $\underset{\sim}{\delta}^0$ selects component $-\tau + 1$; at $t_{(1)}$, $\underset{\sim}{\delta}^1$ component $-\tau + 2$ and so on up to component 0.

2.3.2 Division

Three schemes have been used.

(i)   Trial and error subtraction.

(ii)   Non-restoring subtract and add.

(iii) Successive approximation by subdivision of intervals.

(i) is the familiar method of division, which is not generally favoured because a comparison of current dividend and divisor is needed before every subtraction, (ii) is the most common since the only testing involved, is at the end of a subtraction or addition when the next step is decided. It is also easily adapted to signed division, and (iii) can be shown to reduce to (i).

Since it is intended to control division by the same mechanism as multiplication (to treat it as the reverse process) method (i) will be used. The comparison will be shown to present no problem, since it proceeds in tandem with the subtraction and is controlled by the same t and q counts. A further advantage is that the remainder is always positive without the additional correction step required in method (ii).

The method is explained for the case of a 2τ dimensional integer dividend $\underset{\sim}{d}$ and a τ dimensional divisor $\underset{\sim}{c}$, where $d < c2^\tau$. The quotient is formed in $\underset{\sim}{b}$.

Successive cycles, denoted as before by r, will this time run from $r = 0$ to $r = \tau$ because the dephasing of the q count takes place in the middle of the cycle, instead of at the beginning as in multiplication. Thus initially $r = 0$ for $t = q = 0, 1 \ldots \tau - 1$. The q count is then double counted at $q_{(\tau - 1)}$ and again at $q_{(\tau - 1)}$ in each successive cycle. Exit is again at $q_{(2\tau - 1)} = t_{(\tau - 1)}$ and synchronism is retained.

Comparison takes place at $q_{(0)}$, $q_{(1)} \cdots q_{(\tau - 1)}$. The divider is admitted to the subtractor (if the comparison indicates positive remainder) at $q_{(\tau)}$, $q_{(\tau+1)}$ $\cdots$ $q_{(2\tau-1)}$. However, the comparison must also include the next more significant component of the dividend beyond the range of the divisor since this may be non-zero.

The divisor $\underset{\sim}{c}$ is again synchronized to the q count by double shifting whenever the latter is double counted (at $q_{(\tau-1)}$). Components are set into the least significant end of $\underset{\sim}{b}$ at $q_{(\tau)}$ to form the quotient.

The program is given in Table 2.3.2. Apart from the comparison, the expression is the exact reverse of that given in Table 2.3.1 (multiplication).

Table 2.3.2 Division b = d/c.   ISE

P_0 $\quad (\underset{\sim}{d} \wedge \underset{\sim}{\varepsilon}_{t+\tau-1} \geqslant \underset{\sim}{c} \wedge \underset{\sim}{\varepsilon}_{\tau} | q)_{q_{(0)}, q_{(1)} \cdots q_{(\tau-1)}} \quad \vee \quad (\underset{\sim}{d} \wedge \underset{\sim}{\varepsilon}_{t+\tau})_{q_{(\tau-1)}} \quad \rightarrow$

$\quad \rightarrow (\underset{\sim}{d} \leftarrow \underset{\sim}{d} \wedge \underset{\sim}{\varepsilon}_t - \underset{\sim}{c} \wedge \underset{\sim}{\varepsilon}_{\tau} | q)_{q_{(\tau)}, q_{(\tau+1)} \cdots q_{(2\tau-1)}} \quad \rightarrow (\underset{\sim}{b} \leftarrow \underset{\sim}{\varepsilon}_t)_{q_{(\tau)}}$

$\quad (q \leftarrow q + 1)_{q_{(\tau-1)}}$

P_1 EXIT $\ldots\ldots\ldots\ldots\ldots\ldots\ldots\ldots\ldots\ldots\ldots\ldots\ldots\ldots\ldots\ldots\ldots q_{(2\tau-1)}$

The following notes are added.

(i) In line 1 components are selected from $\underset{\sim}{d}$ by $\underset{\sim}{\varepsilon}_{t+\tau-1}$ and $\underset{\sim}{\varepsilon}_{t+\tau}$. Input to the comparator is thus from stages $\tau - 1$ and τ of the shift register (D) which holds the dividend.

(ii) $\underset{\sim}{d} \wedge \underset{\sim}{\varepsilon}_{t+\tau-1}$ accesses $\underset{\sim}{d}$ components

$\qquad \tau - 1, \tau, \ldots 2(\tau - 1)$ at $q_{(0)}, q_{(1)} \cdots q_{(\tau-1)}$ when r = 0

$\qquad \tau - 2, \tau - 1 \ldots 2\tau - 3$ $\qquad\qquad\qquad\qquad\qquad$ r = 1

$\qquad 0 \ldots\ldots\ldots \tau - 1$ $\qquad\qquad\qquad\qquad\qquad$ r = $\tau - 1$.

(iii) $(\underset{\sim}{d} \wedge \underset{\sim}{\varepsilon}_{t+\tau}) \, q_{(\tau-1)}$ accesses d components

$\qquad 2\tau - 1$ when r = 0

$\qquad 2\tau - 2$ " r = 1

$\qquad \cdots\cdots\cdots\cdots$

$\qquad \tau$ " r = $\tau - 1$.

2.3.3 BCD-binary conversion

The decimal digits of a number in BCD are shown in Fig 2.3.1 as a one-dimensional array of four component vectors. The example shows a 6-digit

integer for which the elements of the array $a^i (i = 5, 4, 3, 2, 1, 0)$ have
weights $10^5, 10^4, 10^3, 10^2, 10^1, 10^0$.

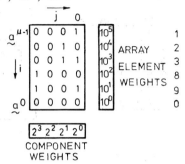

Fig 2.3.1   Example of matrix A holding the number 123 890 in BCD form.

The process used for evaluating the number is expressed as follows.
$$a = (\ldots((a^{\mu-1}10 + a^{\mu-2})10 + a^{\mu-3})10 + \ldots)10 + a^o.$$
This form is suitable for the BCD-binary conversion which will now be
described.

Integer conversion

The central mechanism consists of 2 alternate program steps P_1 and P_2.
In P_1 an accumulator c, initially cleared, is multiplied by 10. In P_2 successive array elements a^i (Fig 2.3.1) are added to c, starting from $a^{\mu-1}$
and finishing with a^o after which the cycle is ended.

The serialisation of the array elements has already been described in
2.2.4.

It remains to explain the transformation of c into 10c. Using the above
notation $c_{\pm r}$ for $2^{\mp r}c$ a multiplication by 10 results from
$$c \leftarrow (c_{-2} + c)_{-1}.$$

The multiplication by 2 outside the bracket is obtained by suppressing
1 shift at $t_{(0)}$. The scaling inside the bracket cannot be achieved by
suppressing shifts since both c and c_{-2} must be available for addition. It
is therefore necessary to extend the shift register at the least significant
end to give the two additional components c_{-1} and c_{-2}.

Fig 2.3.2a illustrates the mechanism for a 5-component vector in which
3 is transformed into 30. Fig 2.3.2b shows the shift register and gating
which has to be applied in P_1 for the multiplication by 10. The shifting
pulses are applied at the clock input as normal. $\overline{t_{(0)} \wedge P_1}$ is applied at
the shift enable input in order to suppress a shift at $t_{(0)}$. The addend

from c_{-2} is introduced into an adder during P_1. Actual circuitry is given in 3.4.2.

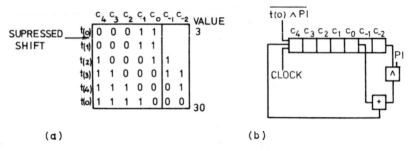

		c_4	c_3	c_2	c_1	c_0	c_{-1}	c_{-2}	VALUE
SUPRESSED	$t_{(0)}$	0	0	0	1	1			3
SHIFT	$t_{(1)}$	0	0	0	1	1			
	$t_{(2)}$	1	0	0	0	1	1		
	$t_{(3)}$	1	1	0	0	0	1	1	
	$t_{(4)}$	1	1	1	0	0	0	1	
	$t_{(0)}$	1	1	1	1	0	0	0	30

(a)   (b)

Fig 2.3.2   Example of multiplication by 10.

The complete program for BCD - Binary Integer Conversion is given in Table 2.3.3.

Table 2.3.3   Program for BCD - Binary Integer Conversion.

		ISE	OSE
P_0	$\underset{\sim}{c} \leftarrow \underset{\sim}{0}, \quad i \leftarrow \mu$		
P_1	$\underset{\sim}{c} \leftarrow (\underset{\sim}{c}_{-2} + \underset{\sim}{c})_{-1}, \quad i \leftarrow i - 1$		$P_2 \wedge \overline{(i = 0)}$
P_2	$\underset{\sim}{c} \leftarrow \underset{\sim}{c} + \underset{\sim}{a}^i$		
P_3	EXIT	$(i = 0)$	

The following explanations are added.

(i)   The operations are given as either operations on vectors or values, whichever is the more convenient.

(ii)   In P_0, $\underset{\sim}{c}$ (accumulator) and suffix i (decimal digit count) are initialized. The latter is set to μ and is decremented in P_1 before the decimal digit $\underset{\sim}{a}^i$ is added.

(iii) The condition $(i = 0)$ determines the end of the conversion. If $\overline{(i = 0)}$, P_1 and P_2 are repeated to take in the next decimal digit. If $(i = 0)$ the least significant digit (the unit decade) has been added and the next step is P_3.

(iv)   The BCD data need not necessarily be available in the parallel form of Fig 2.3.1. It may be provided as a sequence of decimal digits (such as from a tape reader). In this case P_2 would be enabled only when the next decimal digit is available. There must be an interlock with the mechanism which supplies the data. Interlocks are dealt with in 2.5.

Fraction conversion

The conversion of a BCD fraction of say 6 decimal places into binary form requires a matrix of constants E, in which $e^i = 10^{-i}$. Array elements of this matrix are multiplied by the decimal digits of the fraction to be converted and added into an accumulator C.

The matrix E is given in Fig 2.3.3 for a 20-component result $j = 0$, $-1 \ldots, -\nu + 1$ where $\nu = 20$ (the reason for including components -20, -21, -22 will become obvious later).

i \ j	-1	-2	-3	-4	-5	-6	-7	-8	-9	-10	-11	-12	-13	-14	-15	-16	-17	-18	-19	-20	-21	-22	
-1	0	0	0	1	1	0	0	1	1	0	0	1	1	0	0	1	1	0	0	1	1	0	10^{-1}
-2	0	0	0	0	0	0	1	0	1	0	0	0	1	1	1	1	0	1	0	1	1	1	10^{-2}
-3	0	0	0	0	0	0	0	0	0	1	0	0	0	0	0	1	1	0	0	0	1	0	10^{-3}
-4	0	0	0	0	0	0	0	0	0	0	0	0	0	1	1	0	1	0	0	0	1	1	10^{-4}
-5	0	0	0	0	0	0	0	0	0	0	0	0	0	0	0	0	1	0	1	0	1	0	10^{-5}
μ -6	0	0	0	0	0	0	0	0	0	0	0	0	0	0	0	0	0	0	0	1	0	0	10^{-6}

Fig 2.3.3 Constant matrix E, where $e^i = 10^{-i}$.

Negative component suffixes are used and the selection vector is δ_t where $\delta_t^k(\tau) = \varepsilon^{k-\tau+1}(\tau)$. δ_t is again omitted from the notation and $c_{\pm r}$ used in place of $c \wedge \delta_{t\pm r}$.

Before proceeding to give the conversion program, namely the multiplication of E by the fraction to be converted, a simplification of E is explained.

A matrix of constants need never be stored as such. Only components which are 1 need be stored since all others must be 0. Thus E may be stored in the form of a single row vector η of dimension 19 whose components have suffices from -4 to -22 inclusive (see Fig 2.3.3). The components are formed from the Boolean $(i = -1)$, $(i = -2)$, $\ldots$, $(i = -6)$ or disjunctions of these as demanded by the presence of 1's in E. Writing the value of i for the Boolean i equals the value, η is defined by:

$$\eta_{-4} \Rightarrow -1$$
$$\eta_{-5} \Rightarrow -1$$
$$\eta_{-6} \Rightarrow 0$$
$$\eta_{-7} \Rightarrow -2$$
$$\eta_{-8} \Rightarrow -1$$
$$\eta_{-9} \Rightarrow -1 \vee -2$$

$$\eta_{-10} \Rightarrow {}^{-3}$$
$$\eta_{-11} \Rightarrow 0$$
$$\eta_{-12} \Rightarrow {}^{-1}$$
$$\eta_{-13} \Rightarrow {}^{-1} \vee {}^{-2}$$
$$\eta_{-14} \Rightarrow {}^{-2} \vee {}^{-4}$$
$$\eta_{-15} \Rightarrow {}^{-2} \vee {}^{-4}$$
$$\eta_{-16} \Rightarrow {}^{-1} \vee {}^{-2} \vee {}^{-3}$$
$$\eta_{-17} \Rightarrow {}^{-1} \vee {}^{-3} \vee {}^{-4} \vee {}^{-5}$$
$$\eta_{-18} \Rightarrow {}^{-2}$$
$$\eta_{-19} \Rightarrow {}^{-5}$$
$$\eta_{-20} \Rightarrow {}^{-1} \vee {}^{-2} \vee {}^{-6}$$
$$\eta_{-21} \Rightarrow {}^{-1} \vee {}^{-2} \vee {}^{-3} \vee {}^{-4} \vee {}^{-5}$$
$$\eta_{-22} \Rightarrow {}^{-2} \vee {}^{-4}$$

The serialization of E may be realised easily in practice. The components $\eta_{-4} \ldots \eta_{-22}$ are formed by OR gates whose inputs are the Booleans given from the right hand side of the above table. Each component is then selected using a single stage 19-input multiplexer (as encompassed for instance by three 3-bit multiplexers).

Fig 2.3.4 shows an example of a BCD fraction (0.512 389) held in parallel form as the matrix B^i_j. Table 2.3.4 gives the program for its conversion into binary form.

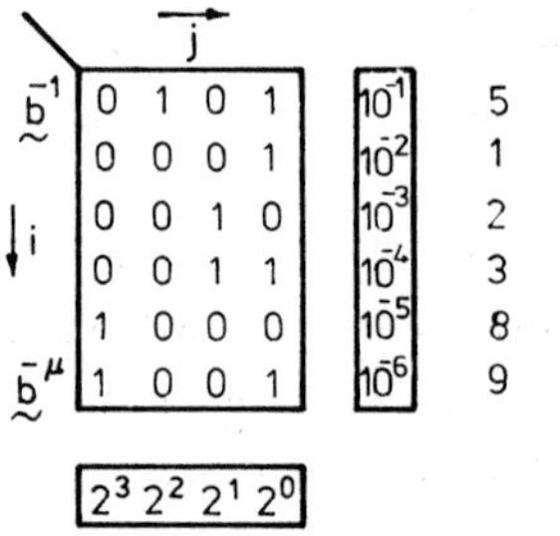

Fig 2.3.4   Example of matrix B holding the BCD fraction 0.512 389.

Table 2.3.4   BCD - Binary Fraction Conversion.

		ISE	OSE
P_0	$c \leftarrow 0,\ i \leftarrow 0,\ j \leftarrow 0$		
P_1	$i \leftarrow i - 1$		$P_2 \wedge (j = 0) \wedge (i > -\mu)$
P_2	$c \leftarrow [(c + \eta_{-j}) \wedge b^i_j],\ j \leftarrow 4\,|\,(j + 1)$		
P_3	EXIT		$(j = 0) \wedge \overline{(i > -\mu)}$

The following explanations are added.

(i) P_2 adds to $\underset{\sim}{c}$ a 20-component vector made up from

$\eta_{-19} \ldots \eta_0$ when $j = 0$

$\eta_{-20} \ldots \eta_{-1}$ when $j = 1$

$\eta_{-21} \ldots \eta_{-2}$ when $j = 2$

$\eta_{-22} \ldots \eta_{-3}$ when $j = 3$ with the assumption $\eta_0 = \eta_{-1} = \eta_{-2} = \eta_{-3} = 0$.

This accounts for the existence of η_{-20}, η_{-21} and η_{-22} which are needed to obtain a result which is accurate to 20 binary places.

(ii) The operation described in (i) is conditional on b_j^i (cf 2.2.3). This gives the effect of multiplication by B.

(iii) j is held in a modulo 4 counter and shown counted in P_2. In practice, it would be counted at the end of P_2, when the termination (j = 3) presets the change of program step (to P_1 if $(i > -\mu)$, to P_3 otherwise).

(iv) BCD data in serial form would require enabling P_2 only after the decimal digit is available.

2.3.4 Binary - BCD conversion

There is a similarity between binary-BCD integer conversion and BCD-binary fraction conversion and again between binary-BCD fraction conversion and BCD-binary integer conversion. A matrix of constants is required in the former but not in the latter. This matrix of constants was used as a multiplicand in BCD-binary fraction conversion and a similar matrix will be used as divisor in binary-BCD integer conversion.

Integer conversion

A binary integer $\underset{\sim}{c}$, held in a shift register, is to be converted into the BCD array $\underset{\sim}{a}^i$ (matrix A) of Fig 2.3.1. The form of storage of $\underset{\sim}{a}^i$ in this case will be conveniently a set of 4-bit counters .

A constant matrix D (Fig 2.3.5) in which $d^i = 10^i$ will be used for the conversion. If the case of a 6-decimal digit integer is treated, $\mu = 6$ (i = 5, 4 ...., 0), while $\nu = 17$, since $10^5 < 2^{17}$.

Starting from $i = \mu - 1$, d^i (in the example 10^5) is subtracted repeatedly from c until the remainder is smaller than d^i. The resulting count $\underset{\sim}{a}^i$ is the coefficient of 10^i. The process is repeated thereafter with the remainder for $i = \mu - 2, \mu - 3 \ldots 0$.

i \ j	16	15	14	13	12	11	10	9	8	7	6	5	4	3	2	1	0	
5	1	1	0	0	0	0	1	1	0	1	0	1	0	0	0	0	0	10^5
4	0	0	0	1	0	0	1	1	1	0	0	0	1	0	0	0	0	10^4
3	0	0	0	0	0	0	0	1	1	1	1	1	0	1	0	0	0	10^3
2	0	0	0	0	0	0	0	0	0	1	1	0	0	1	0	0		10^2
1	0	0	0	0	0	0	0	0	0	0	0	0	0	1	0	1	0	10^1
0	0	0	0	0	0	0	0	0	0	0	0	0	0	0	0	0	1	10^0

$$2^{16}\ 2^{15}\ 2^{14}\ 2^{13}\ 2^{12}\ 2^{11}\ 2^{10}\ 2^9\ 2^8\ 2^7\ 2^6\ 2^5\ 2^4\ 2^3\ 2^2\ 2^1\ 2^0$$

Fig 2.3.5 Constant matrix D, where $d^i = 10^i$.

Again the constant matrix D is not stored as such. An equivalent vector ζ is defined as in the last section by

$$\zeta_{16}, \ \zeta_{15} \Rightarrow 5$$
$$\zeta_{13}, \ \zeta_4 \Rightarrow 4$$
$$\zeta_{10} \Rightarrow 5 \vee 4$$
$$\zeta_9 \Rightarrow 5 \vee 4 \vee 3$$
$$\zeta_8 \Rightarrow 4 \vee 3$$
$$\zeta_7 \Rightarrow 5 \vee 3$$
$$\zeta_6 \Rightarrow 3 \vee 2$$
$$\zeta_5 \Rightarrow 5 \vee 3 \vee 2$$
$$\zeta_3 \Rightarrow 3 \vee 1$$
$$\zeta_2 \Rightarrow 2$$
$$\zeta_1 \Rightarrow 1$$
$$\zeta_0 \Rightarrow 0$$

The complete program is given in Table 2.3.5.

Table 2.3.5 Binary - BCD integer conversion.

		ISE	OSE
P_0	$A \leftarrow 0, \ i \leftarrow \mu - 1$		
P_1	$\beta \leftarrow (\underset{\sim}{c} \geqslant \underset{\sim}{\zeta})$		$P_2 \vee [P_3 \wedge \overline{(c = 0)}]$
P_2	$\underset{\sim}{c} \leftarrow \underset{\sim}{c} - \underset{\sim}{\zeta}, \ a^i \leftarrow a^i + 1$	β	
P_3	$i \leftarrow i - 1$		$P_1 \wedge \bar{\beta}$
P_4	EXIT	$(c = 0)$	

In P_1 a relational operation uses the method described in 2.2.1. The result (Boolean) presets the program to branch to either P_2 or P_3.

Fraction conversion

If the fraction is multiplied by 10, the first decade (after the decimal point) is converted into an integer. Removing the integer the process is repeated with the remainder. The number of repetitions depends on the number of significant decimal digits in the original fraction. For an unbiased round-off half the least significant digit is added to the fraction before conversion (eg a fraction of 6 decimal digits has 5×10^{-7} added).

As in BCD-binary integer conversion multiplication of the shift register content $\underset{\sim}{c}$ by 10 is obtained from

$$\underset{\sim}{c} \leftarrow (\underset{\sim}{c} + \underset{\sim}{c}_{-2})_{-1}$$

Thus 2 additional stages are required at the least significant end of the shift register. Moreover to allow carry propagation in the integer formation, 3 extra stages are needed at the most significant end so that the integer is formed in c_3, c_2, c_1, c_0. A shift register with 3 extra stages necessitates 3 additional shifts per cycle. But one of these is suppressed in the above multiplication by 10. Thus 2 additional shifts at $t_{(0)}$ have the desired effect and the transformation used is

$$\underset{\sim}{c} \leftarrow (\underset{\sim}{c} + \underset{\sim}{c}_{-2})_2 .$$

As a simple example the rather special case of the conversion of 0.111 ($2^{-1} + 2^{-2} + 2^{-3}$) into 0.875 is demonstrated in Fig 2.3.6.

A four-phase clock (see 3.1.3) is used to obtain the additional shifts at $t_{(0)}$. The normal shift is at the fall of clock 3. The content of the shift register is thus shown in Fig 2.3.6 at clock 0 and additionally at clock 1 and clock 2 during $t_{(0)}$. A carry, which occurs in the first cycle during the x10 transformation, is shown at the right of the table. The decimal digits, held in the encased 4-component vector c_3, c_2, c_1, c_0, are shown in a further column.

Table 2.3.6 gives the program for a fraction of μ significant decimal digits. The result is formed in the array $\underset{\sim}{a}^i$ ($i = -1, -2, \ldots -\mu$), which in this case is simply in the form of a set of latching bistables. A masking vector $\underset{\sim}{\alpha}^4$, such as was used in the modulo operation (2.2.2), isolates the integer components c_3, c_2, c_1, c_0. Its negate removes them from the register, to produce the remainder fraction. (In the first operation of P_2, $\underset{\sim}{c} \wedge \underset{\sim}{\alpha}^4$ is treated as a 4-component vector.)

The above completes the list of derived operations which are described in detail.

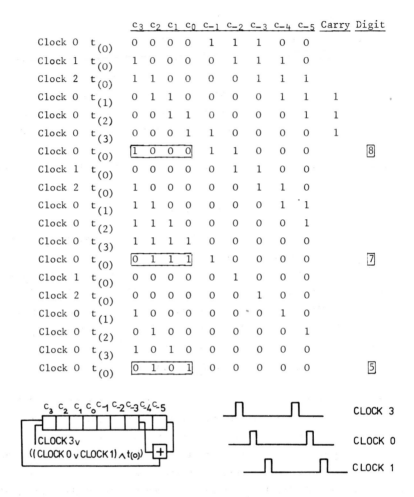

		c_3	c_2	c_1	c_0	c_{-1}	c_{-2}	c_{-3}	c_{-4}	c_{-5}	Carry	Digit
Clock 0	$t_{(0)}$	0	0	0	0	1	1	1	0	0		
Clock 1	$t_{(0)}$	1	0	0	0	0	1	1	1	0		
Clock 2	$t_{(0)}$	1	1	0	0	0	0	1	1	1		
Clock 0	$t_{(1)}$	0	1	1	0	0	0	0	1	1	1	
Clock 0	$t_{(2)}$	0	0	1	1	0	0	0	0	1	1	
Clock 0	$t_{(3)}$	0	0	0	1	1	0	0	0	0	1	
Clock 0	$t_{(0)}$	1	0	0	0	1	1	0	0	0		8
Clock 1	$t_{(0)}$	0	0	0	0	0	1	1	0	0		
Clock 2	$t_{(0)}$	1	0	0	0	0	0	1	1	0		
Clock 0	$t_{(1)}$	1	1	0	0	0	0	0	1	1		
Clock 0	$t_{(2)}$	1	1	1	0	0	0	0	0	1		
Clock 0	$t_{(3)}$	1	1	1	1	0	0	0	0	0		
Clock 0	$t_{(0)}$	0	1	1	1	1	0	0	0	0		7
Clock 1	$t_{(0)}$	0	0	0	0	0	1	0	0	0		
Clock 2	$t_{(0)}$	0	0	0	0	0	0	1	0	0		
Clock 0	$t_{(1)}$	1	0	0	0	0	0	0	1	0		
Clock 0	$t_{(2)}$	0	1	0	0	0	0	0	0	1		
Clock 0	$t_{(3)}$	1	0	1	0	0	0	0	0	0		
Clock 0	$t_{(0)}$	0	1	0	1	0	0	0	0	0		5

Fig 2.3.6 Example of binary-BCD fraction conversion.

Table 2.3.6 Binary-BCD fraction conversion.

		ISE	OSE
P_0	$A \leftarrow 0,\ i \leftarrow 0$		
P_1	$\underset{\sim}{c} \leftarrow (\underset{\sim}{c} + \underset{\sim}{c}_{-2})_2,\ i \leftarrow i - 1$		$P_2 \wedge (i > -\mu)$
P_2	$\underset{\sim}{a}^i \leftarrow \underset{\sim}{c} \wedge \alpha^4,\ \underset{\sim}{c} \leftarrow \underset{\sim}{c} \wedge \overline{\alpha^4}$		
P_3	EXIT	$(\overline{i > -\mu})$	

2.4   OUT OF SEQUENCE ENTRIES

ISEs and OSEs route the program through alternative sequences of operations according to the truth or falseness of given conditions. While the occurrence of conditions and OSEs in the formulation of an algorithm is unrestricted,their links should not be allowed to cross. OSEs may be completely separated,occurring one after the other. If they are not separate, they should be nested, that is occur completely one inside the other. The consequences of violating this rule is illustrated in the example of Table 2.4.1.

Table 2.4.1 Example un-nested OSEs.

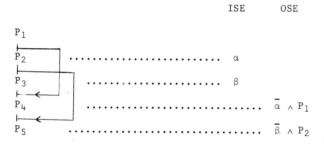

Written as Boolean relations, the result is to shorten the sequence for $\bar{\alpha}$ and $\bar{\beta}$ so that

$$\bar{\alpha} \rightarrow P_1, \ P_4, \ P_5$$

$$\bar{\beta} \rightarrow P_1, \ P_2, \ P_5$$

from which should follow the further shortening to

$$\bar{\alpha} \wedge \bar{\beta} \rightarrow P_1, \ P_5.$$

However, following the OSEs, the result is $\bar{\alpha} \wedge \bar{\beta} \rightarrow P_1, \ P_4, \ P_5$. The OSE due to $\bar{\beta}$ is not followed since the condition β is never encountered.

Logically there is no contradiction since the implication (cf 2.2.1) only assures that an implied operation is not omitted when the condition is true. Thus $\beta \rightarrow P_4$ does not prevent P_4 from being obeyed for $\bar{\beta}$.

The ambiguity is avoided by insisting on the nesting of OSEs. The program would be unambiguous if it were formulated as in Table 2.4.2.

The dummy step P is introduced for clarity but is not needed. The OSE from P_1 could branch according to β such that the OSE to P_4 would follow from $\bar{\alpha} \wedge \beta \wedge P_1$ and to P5 from $(\bar{\alpha} \wedge \bar{\beta} \wedge P_1) \vee (\bar{\beta} \wedge P_2)$.

44

Table 2.4.2 Example of nested OSEs.

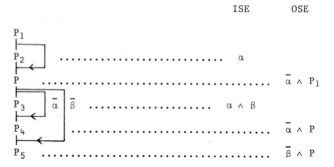

A similar nesting of backward OSEs should be adhered to. However, the restriction on the interlinking of forward and backward OSEs must be stated differently. The sequence of program steps between the condition and the corresponding OSE is called the range of the OSE. The general rule is then to allow a jump from within the range to outside it, but not vice versa*. (Fig 2.4.1 illustrates 4 possible situations.)

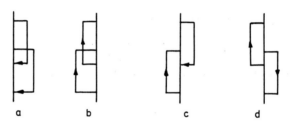

Fig 2.4.1 Four possible interlinkings of forward and backward OSEs.

a and b show the cases of failure to nest forward and backward OSEs respectively. These have been shown to lead to inconsistencies. One of the OSEs is from outside a range to inside it, in each case.

In c both OSEs are from outside to inside and it is certainly to be avoided. d however has both OSEs going from inside a range to outside it and hence is consistent. In fact it is a situation which occurs quite frequently in practical algorithms.

An example is the search for an item in a list. This might be realized as in Table 2.4.3.

---

*In Algol the block structure ensures this. In Fortran the jump into a DO loop is prohibited, but un-nested conditional jumps are allowed.

Table 2.4.3 Search algorithm.

		ISE	OSE
P_1	Set β if required item not found		$\overline{\alpha} \wedge P_3$
P_2	Move to next item in list	β	
P_3	Set α if list exhausted		
P_4	Give indication of failure of search	α	
P_5	Give result of successful search		$\overline{\beta} \wedge P_1$

UNCONDITIONAL
OSE or STOP

2.4.1 Forward OSE

In the above the forward OSE has been used in the form, IF (α) THEN,,,,;
where α is the condition and the ; indicates the end of the conditional statement.

For a large number of cases this form is adequate. It may, for instance,
be used in an algorithm which presets variables before the condition is
encountered. If these variables are then changed during the conditional
statement, when α is true, the result is defined in both cases.

There are odd cases when the form of alternatives IF(α)THEN,,,,ELSE,,,;
is needed. An example is shown in Table 2.4.4.

Table 2.4.4 Example of algorithm using alternative sequences.

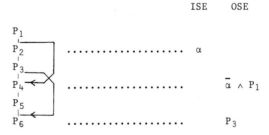

	ISE	OSE
P_1		
P_2		α
P_3		
P_4		$\overline{\alpha} \wedge P_1$
P_5		
P_6		P_3

α implies that program steps P_2 and P_3 are obeyed but not P_4 and P_5. $\overline{\alpha}$
implies the reverse. A second downward pointing arrow crossing the first
shows the unconditional OSE into P_6 from P_3.

A nesting of the above forms is needed where two conditions are involved
say α and β. The four alternatives $\overline{\alpha}\overline{\beta}$, $\alpha\overline{\beta}$, $\overline{\alpha}\beta$ and $\alpha\beta$ are taken into account
in the example of Table 2.4.5.

Table 2.4.5 Algorithm used to discriminate $\overline{\alpha\beta}$, $\alpha\overline{\beta}$, $\overline{\alpha}\beta$ and $\alpha\beta$

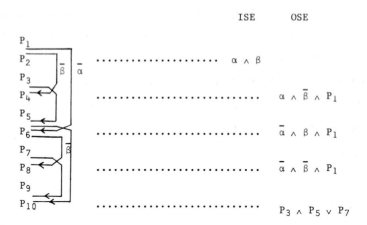

	ISE	OSE
P_1		
P_2	$\cdots\cdots\cdots\cdots\cdots\cdots\cdots\cdots$	$\alpha \wedge \beta$
P_3		
P_4	$\cdots\cdots\cdots\cdots\cdots\cdots\cdots\cdots$	$\alpha \wedge \overline{\beta} \wedge P_1$
P_5		
P_6	$\cdots\cdots\cdots\cdots\cdots\cdots\cdots\cdots$	$\overline{\alpha} \wedge \beta \wedge P_1$
P_7		
P_8	$\cdots\cdots\cdots\cdots\cdots\cdots\cdots\cdots$	$\overline{\alpha} \wedge \overline{\beta} \wedge P_1$
P_9		
P_{10}	$\cdots\cdots\cdots\cdots\cdots\cdots\cdots\cdots$	$P_3 \wedge P_5 \vee P_7$

α implies program steps P_2, P_3, P_4, P_5 but not P_6, P_7, P_8, P_9. $\overline{\alpha}$ implies the reverse. β implies P_2, P_3 or P_6, P_7 but not P_4, P_5 and P_8, P_9. $\overline{\beta}$ implies P_4, P_5 or P_8, P_9 but not P_2, P_3 and P_6, P_7.

The nesting of conditional statement has been observed, so that the result of all program steps is defined. There is no case where this rule cannot be observed. Where a difficulty appears, it can always be resolved by negating a condition so as to exchange the OSE and ISE.

2.4.2 Backward OSE

A backward OSE is used for the control of a loop (DO or FOR statement in computer languages). Only the IF(α)THEN,,,; form can occur since the program must return to the point from which the backward OSE is made or pass beyond it.

Since a loop is involved, there are only two ways of leaving it. Either

(i) within the range of the backward OSE a change occurs, such that the condition for the OSE changes from true to false, or

(ii) a forward OSE occurs within the loop. It takes the program to a step outside the loop.

These alternatives are illustrated in the two equivalent examples of Table 2.4.6.

Table 2.4.6 Control of loops, 2 alternative methods.

(i)   ISE   OSE

$$\begin{array}{ll} P_1 & i \leftarrow 0 \\ P_2 & i \leftarrow i + 1 \\ P_3 \\ P_4 \\ P_5 \end{array}$$

ISE column / OSE column:

P₂ row: OSE: $(i < n) \wedge P_4$

P₅ row: ISE: $\overline{(i < n)}$

(ii)

$$\begin{array}{ll} P_1 & i \leftarrow 0 \\ P_2 & i \leftarrow i + 1 \\ P_3 \\ P_4 \\ P_5 \end{array}$$

P₂ row: OSE: P_4

P₃ row: ISE: $\overline{(i > n)}$

P₅ row: OSE: $(i > n) \wedge P_2$

Program steps P_3 and P_4 are obeyed for $i = 1 \ldots n$ and the sequence from P_5 onwards after $i = n$. At P_5, $i = n$ in the first scheme, while $i = n + 1$ in the second scheme. On the other hand the first scheme gives the wrong result for $n = 0$.

A third possibility is given in Table 2.4.7.

Table 2.4.7 Control of loop, further alternative.

                                     ISE   OSE

$$\begin{array}{ll} P_1 & i \leftarrow 0 \\ P_2 & i \leftarrow i + 1 \\ P_3 \\ P_4 \\ P_5 \end{array}$$

P₂ row: ISE: $(i < n)$ OSE: $(i < n) \wedge P_4$

P₅ row: ISE: $\overline{(i < n)}$ OSE: $\overline{(i < n)} \wedge P_1$

The right action is taken when $n = 0$, and $i = n$ at P_5.

2.5 INTERLOCKS

In a decentralised controller a number of programming units can function autonomously and simultaneously. This does not mean that they are necessarily independent. Their interaction has to be taken into account by a system of interlocks.

It is proposed to control all interactions solely by the interlocking of program controllers. Several situations can arise, and these will be illustrated in the following example.

A programming unit P prepares data for use by another programming unit Q. At a certain point, Q cannot proceed until the data is ready. Therefore, the end of the program step (in P) which forms the data must 'permit' the program step (in Q) which uses the data. The data itself is in a common store accessible to both programs.

A more efficient arrangement is for the data to be held in a double store. One of these stores is available for loading by P, while the other controls actions in Q. The stores alternate between these functions. A changeover takes place when the old data is no longer required by Q and the new data has been fully loaded by P. Two cases arise

(i)   The P cycle is slower than the Q cycle. Then changeover is then a program step in Q. It is 'permitted' by the end of the program step in P which forms the data.

(ii)   The Q cycle is slower than the P cycle. The changeover is a program step in P. It is permitted by the end of the program steps in Q which have been using the data loaded in the previous P cycle.

Frequently what is required is the mutual 'permitting' of a program step in P and one in Q. The two steps are then obliged to commence simultaneously. This would be needed where, due to variations, both cases i and ii, described above, can occur.

A mutual interlock also assures that there is a one to one correspondence between cycles of P and Q. Without it, one part of the control unit would have no means of knowing whether a program step in another has been reached after one or more than one complete cycle.

In the case of two programming units using a common store, one or both may alter the store. The program steps involved might then have to be subject to a mutual inhibition. This has the effect of serializing the calls on the common store, while leaving the order of the program steps free.

Interlocking implies a potential idling in one or other of the interlocked programming units. This means allowing a unit to idle in any program step if required.

Suppose P_n is a program step in P during which a certain action is intended (it could be a computation or the energising of an actuator of some kind). When P enters state P_n no action is taken unless and until the enable

P_n (E_{P_n}) signal becomes true. Once enabled, the termination of P_n is either

(i)   the end of the computation as indicated by a timing signal, in which case this terminating signal must become effective only after E_{P_n} has become true, or

(ii) a direct consequence of the action due to P_n such as a microswitch or other sensor output, which can only occur after the action is enabled.

Through the enabling signals the correct sequencing and relative phasing of all activities can be achieved, irrespective of their duration. In the ideal control system the sequencing of operations is time independent.

However, in deciding on the form of interlocks, one distinction should be made at the outset. It matters whether variations in the duration of operations are inherent in the process or whether they are due only to exceptional circumstances, such as the malfunctioning of a mechanical element. In the former case the mutual interlocking of program controllers should be loose, so that only the overlap of a sequence of steps from each controller is required. This avoids unnecessary restrictions on the process. Time lost in one cycle of a program may be regained in a subsequent cycle or vice versa. If durations are basically fixed, nothing is lost by imposing a mutual interlock of a definite program step from each controller, to give rigid synchronisation.

In the following, three forms of interlocks will be described: Permit, Mutual Permit and Inhibit. All forms of interaction between program units can be handled by these.

2.5.1   Permit

An undirectional 'Permit', say Q_m permits P_n, requires Q_{m-1} to be completed before P_n becomes active. Strictly speaking, such a 'Permit' can occur in isolation (unaccompanied by a reverse Permit) only if the programming unit Q is not cyclic. Otherwise there would be no meaning to before and after Q_m.

In this simple case the enable (E_{P_n}) is just the state of a bistable set by Q_m and reset by P_{n+1}. The action under P_n is permitted by $E_{P_n} \wedge P_n$.

2.5.2   Mutual permit

The simpler case of a mutual 'Permit' by a pair of program steps, which is taken first, is Q_m permits P_n and P_n permits Q_m. The reciprocal enable E_{P_n, Q_m} must be symmetrical in P_n and Q_m. Written without suffices it is defined by

$$E = [E \wedge (P_n \vee Q_m)] \vee (P_n \wedge Q_m).$$

Action under P_n is permitted by $E \wedge P_n$, action under Q_m by $E \wedge Q_m$. Fig. 2.5.1 shows timing diagrams for (a) Q_m entered before P_n, (b) Q_m entered after P_n.

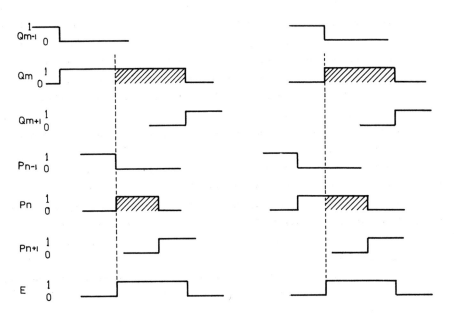

(a) Q_m before P_n (b) Q_m after P_n

Fig 2.5.1 Mutual permit of Q_m and P_n.

The active periods ($Q_m = 5$, $P_n = 3$) are shown shaded. The vertical broken lines show the point in time at which the cycles P and Q are locked. At this point E is set and remains true as long as the disjunction of Q_m and P_n.

E can become successively true only after complete cycles of P and Q, but there is one exception. If P is held up at P_n while Q completes a cycle, E would still be true when Q returns to Q_m, or the reverse situation when Q_m is held up. Hence for a mutual interlock to be completely effective, P_{n-1} and Q_{m-1} must be enabled by $\overline{E}$. Other permits of Q by P and P by Q may of course accompany the mutual permit, in which case the one to one correspondence of cycles is already guaranteed.

The extension to three or more programming units follows immediately. Fig 2.5.2 shows a typical case of three programming units P, Q, R which are locked at P_n, Q_m and R_l by the enable $E = \left[E \wedge (P_n \vee Q_m \vee R_l) \right] \vee (P_n \wedge Q_m \wedge R_l)$ and where Q_{y-1} must be completed before P_w, P_{r-1} before R_s and R_{x-1} before Q_v. Clearly the one to one correspondence of cycles in P, Q and R is guaranteed.

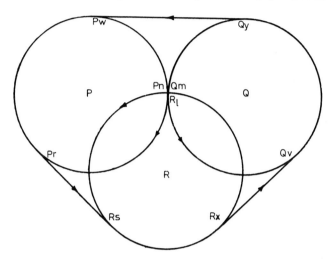

Fig 2.5.2 Diagrammatic representation of three interlocked program units.

The other, less tight, way of locking cycles is to take a sequence of program steps from each and ask for at least one overlap.

Let the sequences be Q_m, Q_{m+1} ... Q_{m+r} and P_n, P_{n+1} ... P_{n+s}. Writing $Q_{m,r}$ for $Q_m \vee Q_{m+1} \vee ... \vee Q_{m+r}$ and $P_{n,s}$ for $P_n \vee P_{n+1} \vee ... \vee P_{n+s}$. Two enables are defined.

$$E_1 = (E_1 \wedge Q_{m,r}) \vee (Q_{m,r} \wedge P_n) \quad \text{and} \quad E_2 = (E_2 \wedge P_{n,s}) \vee (P_{n,s} \wedge Q_m)$$

which are respectively true for the excess Q sequence if P_n is after Q_m and for the excess P sequence if Q_m is after P_n. Then

$$(E_1 \vee E_2) \wedge Q_{m+r}, \quad \bar{E}_1 \wedge P_{n-1}$$

assures that the Q sequence is not completed before the P sequence starts and that there will only be one P sequence for every Q sequence. Similarly

$$(E_1 \vee E_2) \wedge P_{n+s}, \quad \bar{E}_2 \wedge Q_{m-1}$$

delays the completion of the P sequence if required. Clearly the case

$r = s = 0$, when $E_1 \vee E_2 = E$, reduces to the call for simultaneous Q_m and P_n as in the first type of interlock.

Another extension to the setting of a mutual enable E_{P_n}, Q_m is needed where the P and Q program units are sited in controllers which are remote from one another. The mutual enable of P_n and Q_m must occur in synchronism with the Characteristic Period at each controller. To assure this, it is necessary to duplicate the enable, so that a separate bistable is set at each controller. This only requires the transmission of P_n to Q and Q_m to P. $P_n \wedge Q_m$ sets up both (E_{P_n}, Q_m) P and (E_{P_n}, Q_m) Q and these are reset by $\overline{P}_n$ and $\overline{Q}_m$ respectively.

2.5.3   Inhibit

The purpose of this interlock is to prevent an overlap of program steps Q_m and P_n.

An enable Q_m is defined as

$$E_{Q_m} = Q_m \wedge (\overline{P}_n \vee E_{Q_m}).$$

An enable P_n as

$$E_{P_n} = P_n \wedge (\overline{Q}_m \vee E_{P_n}).$$

Whichever program step occurs first, is allowed to run to completion, before the other program step becomes active. In fact the enables are mutual complements so that $E_{Q_m} = Q_m \wedge \overline{E_{P_n}}$ and $E_{P_n} = P_n \wedge \overline{E_{Q_m}}$, as can be easily seen from the above definitions. (Fig 2.5.3 shows timing diagrams similar to those of Fig 2.5.1.)

2.5.4   Commoning of program controllers

The above formulations of interlocks are free from any quantitative element. They impose no restrictions on the duration of program steps.

However it would be wrong to conclude that all the required interlocks between mechanisms can be reduced to the logical interlocks described above. Proportional interlocks, which may take the form of mechanical or fluid linkages or servo loops, are often an essential part of the mechanism. Alternatively two separately controlled mechanisms may depend on one another because of free moving components passing between them.

The durations and occurrence of the corresponding program steps are then no longer independent and the program control should allow for this.

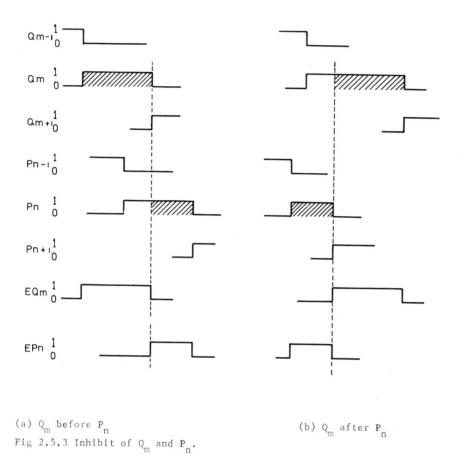

(a) Q_m before P_n (b) Q_m after P_n

Fig 2.5.3 Inhibit of Q_m and P_n.

The objective of replacing mechanical linkages by program interlocks is generally desirable, since it frees the construction of the mechanism from mechanical constraints. However the advantages are not wholly one-sided.

An example is a dispensing mechanism, program-controlled in one cycle, which must be emptied and retracted to receive a part released into it in another cycle. If the releasing mechanism is not mechanically linked to the dispenser, the only way of interlocking the operations by program interlock, would be for the release to be permitted by the end of the retraction of the dispenser. The additional delay may be unacceptable and make the linkage necessary.

If such a linkage exists between mechanisms which occur in otherwise separate cycles, the logical organisation of the program control should bring the programs together at a common program step.

Fig 2.5.4 illustrates the sequence for two program controllers P and Q. P_{m-1} and Q_{n-1} terminated respectively by p_{m-1} and q_{n-1} are the program steps preceding the common step R. R is set by $p_{m-1} \vee q_{n-1}$ and an enable bistable E (which is normally in a true state) is

reset by $(p_{m-1} \vee q_{n-1}) \wedge \bar{R}$, and

set by $(p_{m-1} \vee q_{n-1}) \wedge R$.

E $\wedge$ R is then the active part of R (shown shaded) and r is the termination, which initiates P_m and Q_n, where the cycles once more separate.

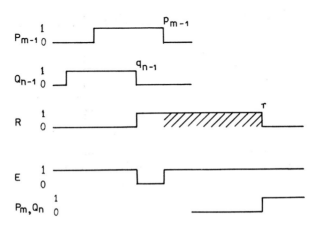

Fig 2.5.4 Commoning of program controllers at a program step for interlocked mechanism.

2.6   SUBPROGRAMS

A program which is subsidiary to and called by the main program is called a subprogram. The term subprogram is used because the form is inter-mediate between that of a microprogram and a subroutine.

It differs from the former since it encompasses more than a sequence of elemental operations. A subprogram can take the form of a normal program with its own registers, bistables and sequence control and can be used at any level in the organisation of the control unit.

It differs from a subroutine in not using the link setting method of entry. This form of organisation is impossible in a decentralised control unit since there is no common index by which to identify the links set by various program controllers.

In the simplest case there is one subprogram which all other programs may call (though only at different times). A single bistable (α) signifies the execution of the subprogram and the suspension of the main program at the program step which called the subprogram.

If separate bistables are associated with each subprogram, several subprograms may be called independently of one another and a subprogram may itself call a subprogram.

One reason for the use of subprograms is to allow the modular construction of control units. The modules could consist of previously designed and proven subprograms. As an example a computational unit may use subprograms for multiplication, division, input/output or other derived operations. Another example is a pulsed drive, where the main program computes the movements. Before passing this data onto the servo, an acceleration/deceleration subprogram might be needed.

Most of these applications involve data or parameter transfer between the main program and the subprogram and vice versa. If the two programs have the same resolution, a common synchronising count can be used and there is no difficulty in this parameter transfer. If, however, the programs are constructed for different resolutions and hence have different Characteristic Periods, there must be a synchronising procedure at the point where control is passed from one program to the other.

2.6.1 Parameter transfer

The simplest case of parameter transfer is when

(i)   the Characteristic Periods of main program and subprogram are the same,
(ii) the period of the main program is an integral multiple of the period of the subprogram.

Fig 2.6.1 illustrates case i, where P_0, P_1, P_2 are 3 consecutive program steps in the main program. In P_0 parameters are transferred into the subprogram stores. In P_2 the results of the subprogram are transferred back to the main stores. In P_1 the subprogram Q_0, Q_1, Q_2 is called.

All program steps are shown with a terminating signal and may be of variable duration. However the subprogram exit step Q_2 must be only one Characteristic Period long.

P_0 is the signal for the entry into the subprogram. Bistable α is also set by p_0 and is subsequently reset by q_1. $\bar{\alpha}$ becomes the terminating signal for P_1. The result is that the end of P_1 coincides with the end of Q_2, whereupon the main program is allowed to continue in P_2.

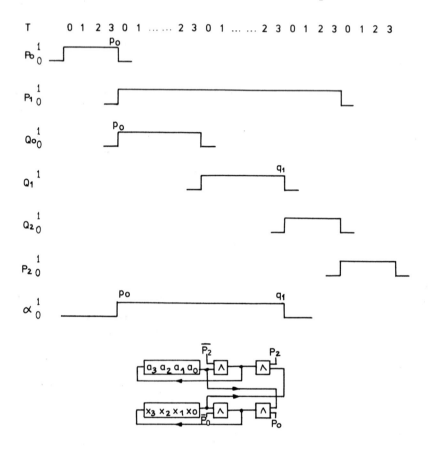

Fig 2.6.1 Parameter transfer for equal resolution programs.

To illustrate the parameter transfer two registers A and X are shown, which belong to the main and subprograms respectively. If the subprogram is described by $x \leftarrow f(x)$, $P_0 \rightarrow (x \leftarrow a)$ and $P_2 \rightarrow (a \leftarrow x)$.

Case ii is unlikely to arise in arithmetic units since the subprogram is normally required to have a resolution greater than that of the calling program (case treated in the next section).

There is however an important instance where the subprogram has a resolution which is a submultiple of the main program. This is the case of a digital servo control where the subprogram performs its function by counting techniques while the main program computes movements by serial arithmetic.

Since the subprogram has a Characteristic Period equal to the clock
period,it can commence immediately it is called. Moreover parameter transfers present no difficulty,since the transfer to and from counters is performed in parallel and takes place within a clock period. The only problem
of synchronizatfon is at the end of the subprogram. The subprogram will in
general terminate before the end of Characteristic Period of the main program.
The bistable α serves to staticise this condition.

2.6.2 Synchronization

The following mnemonics are used in this section.

CP   for Characteristic Period

ECP for end of CP signal, and

SC   for synchronizing count.

A main program calling a subprogram whose CP is longer,than its own,
involves

(i)   an initial alignment of main program and subprogram SCs,

(ii)   a delay after parameter input before the subprogram can commence,

(iii) a freezing of subprogram registers after the subprogram exit step.

Provided the subprogram CP is less than twice the main program CP (not
a serious restriction), an extension of the technique described in the last
section can be used.

Fig 2.6.2 illustrates the case of a 5-bit resolution main program P_0,
P_1, P_2 (SC:S, ECP:S4). As before,P_0 and P_2 are parameter transfer steps.
P_1 calls an 8-bit resolution subprogram Q_0, Q_1, Q_2 (SC:T, ECP:T7).

The subprogram SC(T) is initially zero and will be left in a zero state
after the subprogram exit. The T count starts to count with P_0. Since the
subprogram ECP will then be later than the main program ECP,the change to P_1
occurs before the subprogram is entered. Therefore P_1 itself becomes the
setting signal for Q_0 and α *. The T count continues through P_1, but only to
the end of Q_2.

In addition to α which holds the main program in P_1,a further bistable
β is needed. At the end of Q_2, α is reset and β is set. $\overline{\beta}$ gates the shifting
clock of the subprogram registers. Thus after Q_2, and between the ECPs of sub-
program and main program, their contents are held in origin-position.

---

*If the subprogram CP were greater or equal to twice the main program CP, P_1
would change to P_2 before Q_0 and α are set.

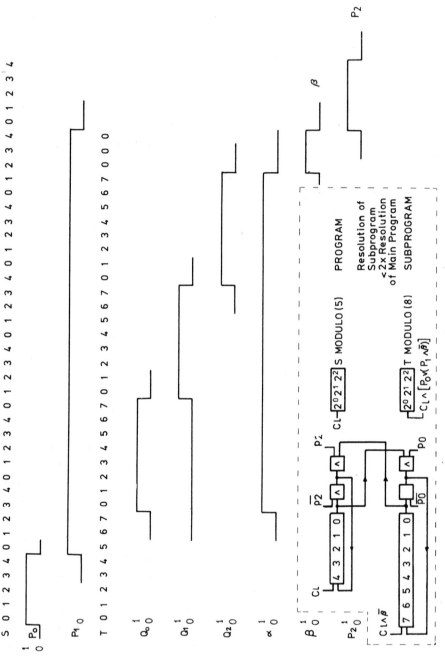

Fig 2.6.2 Parameter transfer for unequal resolution programs.

Since α is reset at the end of Q_2, the first ECP of the main program following it, changes P_1 to P_2. P_2 immediately resets β so that the transfer between subprogram and main program registers (now synchronized) can take place. To avoid separate resetting of the T count, its counting clock is inhibited by $\bar{\beta}$. This has the effect of leaving T in a zero state after the subprogram call.

The main program SC(S) has been unaffected throughout. All main program registers have remained synchronized to it. There is moreover no reason why separate action cannot take place in the main program during P_1.

There is a similar problem of synchronization in the exchange of data between controllers P and Q, if these are remote from one another.

Let Q again be the receiving station. P_m transmits data to be received in Q_n. If Q_n is after P_m, the sending station P will not be notified till Q_n has reached it. Only then can data be transmitted and this itself will have its own propagation delay.

The duplicate interlock system (end of 2.5.2) fails to give synchronization. Since moreover the data rates may be much higher than the rate of change of program steps, the transmission elements may impose quite different propagation delays. It follows that the only satisfactory solution is the use of a buffer store at the receiving station.

P_m can then initiate data transmission without waiting for Q_n. The transmitted data is preceded by a beginning of transmission signal and ended by an end of transmission signal. The former prepares the buffer store and synchronizes its clock in readiness to receive data. The latter freezes the buffer and clock, and sets a bistable which is used in place of P_m in setting up the enable $(E_{P_m, Q_n})_Q$. The buffer clock is restarted by this enable, so that the transfer from buffer store to working registers is again synchronized. The data must be transmitted in fixed format for this mechanism to work.

The extension to the use of a double buffer store (beginning of 2.5) introduces no new problems. Frequently one buffer is sited in P and the other in Q to facilitate the synchronization.

Methods of data transmission for moderate distances are described in 4.2.2. When data links become too long for these techniques to cope, carrier-based telecommunication techniques must be used. In that case format requirements have to be adapted to the needs of telecommunication practice. All too often, however, complex data transmission schemes are used, even for short distances, where simpler methods would do.

An important simplifying factor is the localisation of peripheral control functions, to reduce the quantity and frequency of data interchange.

2.7 DETAILING OF OPERATIONS IN ALGORITHMS

2.7.1 Description of example

The application of the design principles of the previous sections to the design of a special-purpose computing unit will now be illustrated by an example.

The function of the unit is to compute the distance between two points on a workpiece from the two co-ordinate pairs read by an X,Y table. If x_1, y_1 and x_2, y_2 are the co-ordinates, the computation produces

$$d = \{(x_2 - x_1)^2 + (y_2 - y_1)^2\}^{\frac{1}{2}}.$$

The purpose of the unit is to avoid having to align the workpiece, so that the distance to be measured lies along one or other of the co-ordinate axes of the table. The computation must be performed at such a speed that the display of d does not significantly differ from the display of the co-ordinates themselves.

A very simple modification of the unit would allow the retention of x_1, y_1 so that a series of

$$r_i = \{(x_i - x_1)^2 + (y_i - y_1)^2\}^{\frac{1}{2}}$$

could be obtained and thereby allow the exploration of circular features on a workpiece. This avoids the need for a rotary table and the difficulty of making the centre of the feature coincide with the axis of rotation of the rotary table.

The mode of readout from the X, Y table varies, depending on the technique used. It is therefore assumed that the values read out have undergone the necessary transformations and are available in the form of a pair of 6-decimal digit BCD numbers. If the table origin is outside the workpiece, these numbers are positive integral multiples of the smallest discernible increment, say 1 μm. The computation will be performed throughout at this resolution (20 binary places) and the output display will also have 6 decimal digits. The area covered by the measurements is 0.7 m x 0.7 m and no distance from the origin must exceed 1 m.

The circuitry can be built with less than 100 integrated circuit modules. Using the single length Characteristic Period (20 clock periods) as the unit of time, 170 periods on average, complete the operation. If a master clock of 1 MHz is used to produce a two-phase clock of 2 μs period, the total time is of the order of 7 ms.

In outline the computation is as follows.

a   The computing unit idles, waiting for a data strobe (ST) to indicate that readings x_1, y_1 are available at the input.

b   BCD - Binary Conversion Subprogram is entered to input x_1 into $\underset{\sim}{b}$, y_1 into $\underset{\sim}{c}$.

c x_1 is stored in $\underset{\sim}{x}$, y_1 in $\underset{\sim}{y}$ and the unit waits for a further data strobe to indicate that x_2, y_2 are available at the input.

d   BCD - Binary Conversion is entered to input x_2 into $\underset{\sim}{b}$, y_2 into $\underset{\sim}{c}$.

e   Comparisons of x and b, y and c set conditions for subtrahend/diminuend reversal so that positive co-ordinate differences are formed:

$$|x_2 - x_1| \text{ in } \underset{\sim}{x}, \qquad |y_2 - y_1| \text{ in } \underset{\sim}{y}.$$

f $\underset{\sim}{x}$ is set as multiplicand and multiplier and the multiplication subprogram is entered to form the double length product $(x_2 - x_1)^2$ in $\underset{\sim}{z}$.

g   f is repeated for $\underset{\sim}{y}$ and the result added to $\underset{\sim}{z}$ so that

$$z = (x_2 - x_1)^2 + (y_2 - y_1)^2.$$

h   The square root subprogram is entered to form $\sqrt{z}$ in $\underset{\sim}{d}$ (single length).

i   The Binary - BCD Conversion subprogram is entered to convert $\underset{\sim}{d}$ into a form suitable for output or display.

j   The program returns to 'a' in readiness for a new set of readings.

2.7.2 Preliminary considerations

The computation will use the serial techniques described above. Additional choices for this particular computation are:

a   Because of the relatively high resolution (20 binary places) and the time available for the computation, it is preferable to serialize the two multiplications and avoid the need for extra registers.

b   Since two working registers (for $\underset{\sim}{b}$ and $\underset{\sim}{c}$) are required for multiplication, they can be used in BCD - Binary Conversion for x and y data respectively. This allows the inputs to take place in parallel and avoids input multiplexing, since x will always be read into $\underset{\sim}{b}$ and y into $\underset{\sim}{c}$.

c   Since BCD - Binary Conversion (contrary to the reverse process) takes a fixed time, irrespective of the input values, the one subprogram (see 2.3.3) can control input and conversion of both x and y. In the program it is simply necessary to operate on $\underset{\sim}{b}$ (input from x) as well as $\underset{\sim}{c}$ (input from y).

d   While there is some difficulty in transferring control from a single length main program to a double length subprogram (Section 2.6.2),the reverse presents no problems. After the multiplication subprogram (single length) the transfer of the double length result to z involves a double length main program step.

e   After the second multiplication and the double length transfer, the double length subprogram for square root (see below) is entered.

f · Since the square root is followed immediately by Binary - BCD Conversion, there is no need to return to the main program. However the entry into the conversion subprogram is accompanied by a new step in the main program. During this step the circulation of the register, whose content is to be converted, is halved so as to allow the conversion to proceed single-length.

g   The two final subprograms (square root and Binary - BCD Conversion) do not use their own working registers but operate on registers which are drawn from the main program.

h   The basic operations which have to be performed are different for the various registers used. Hence, rather than design each register for the most complex operation and carry redundant circuitry, the operations permitted on each register are derived from the following list and combined according to requirement.

 0   Addition to register including replacement of content
 1   Addition or subtraction
 2   Reversal of subtrahend and diminuend
 3   Additional access points to register (see BCD - Binary Conversion 2.3.3)
 4   Double shift of register during one clock period
 5   Suppression of shift during one clock period
 6   Change from double length to single length circulation and vice versa.

2.7.3 Additional material

Provision for operations 1 and 2 in the above list needs an extension to the algorithm for addition and subtraction.

In 2.2.2 addition and subtraction were given as an expression for the generation of $a \pm b$ and the carry c. Since the expression for the former is symmetrical in a and b and the same for addition and subtraction, it remains unchanged.

The expression for $\underset{\sim}{c}$ can be given a slightly different form:

$$\underset{\sim}{c} = \underset{\sim}{a} \wedge \underset{\sim}{b} \vee [(\underset{\sim}{a} \vee \underset{\sim}{b}) \wedge \underset{\sim}{c}_{-1}] \quad \text{for } a + b$$

$$\underset{\sim}{c} = \bar{\underset{\sim}{a}} \wedge \underset{\sim}{b} \vee [(\bar{\underset{\sim}{a}} \vee \underset{\sim}{b}) \wedge \underset{\sim}{c}_{-1}] \quad \text{for } a - b.$$

It is convenient to use a Boolean Γ to discriminate between addition and subtraction, so that by defining

$$\underset{\sim}{c} = \underset{\sim}{a}' \wedge \underset{\sim}{b} \vee [(\underset{\sim}{a}' \vee \underset{\sim}{b}) \wedge \underset{\sim}{c}_{-1}]$$

and

$$\underset{\sim}{a}' = (\underset{\sim}{a} \wedge \bar{\Gamma}) \vee (\bar{\underset{\sim}{a}} \wedge \Gamma)$$

the conditional operation $\underset{.}{a} + (-1)^{\Gamma}\underset{\sim}{b}$ is provided.

The reversal of subtrahend and diminuend is allowed for by the following definitions

$$\underset{\sim}{c} = \underset{\sim}{a}' \wedge \underset{\sim}{b}* \vee [(\underset{\sim}{a}' \vee \underset{\sim}{b}*) \wedge \underset{\sim}{c}_{-1}]$$

$$\underset{\sim}{a}' = (\underset{\sim}{a}* \wedge \bar{\Gamma}) \vee (\bar{\underset{\sim}{a}*} \wedge \Gamma)$$

$$\underset{\sim}{a}* = (\underset{\sim}{a} \wedge \bar{\gamma}) \vee (\underset{\sim}{b} \wedge \gamma)$$

$$\underset{\sim}{b}* = (\underset{\sim}{b} \wedge \bar{\gamma}) \vee (\underset{\sim}{a} \wedge \gamma).$$

The most general operation is then

$$(-1)^{\gamma\Gamma}a + (-1)^{\bar{\gamma}\Gamma}b.$$

The above shows what additional circuitry must be provided for operations 1 and 2 listed in Subsection 2.7.2. With each additional facility the gating chain is lengthened and the connections to $\underset{\sim}{a}$ and $\underset{\sim}{b}$ brought further back along it.

One derived operation is used in the example, which has not been described in Section 2.3. This is the square root process. It will be given without explanation since it is easily verified with an actual example.

The square root of z is generated in a vector $\underset{\sim}{d}$ which is of the same resolution as $\underset{\sim}{z}$. The resolution must be even (say 2τ). The synchronizing count q is a modulo (2τ) count. It is double counted and $\underset{\sim}{d}$ is double shifted at the end of the first phase of a two-phase clock.

The notation $\underset{\sim}{q}_{(i)}$ is used for the 2τ dimensional vector which has a 1 as the least significant component and 0 elsewhere, when $q = i$.

$$\underset{\sim}{q}_{(i)} = (\underset{\sim}{\varepsilon}^{o})_{q(i)} = (\underset{\sim}{\varepsilon}^{i})_{q(o)}.$$

The subprogram is given in Table 2.7.1.

Table 2.7.1 Square root algorithm.

		ISE	OSE
R_0	$d \leftarrow 0, \; q \leftarrow 0$		
R_1	$d \leftarrow d - q_{(2\tau-1)}$ $\cdots\cdots\cdots\cdots\cdots\cdots\cdots\cdots\cdots$	$[(R_2 \wedge \bar{\delta}) \vee R_3]$ $\wedge \; \overline{q_{(2\tau-1)}}$	
R_2	$(d \leftarrow d_{+1})_{t(o)}, \; (q \leftarrow q + 1)_{t(o),t(1)}, \; \delta \leftarrow (z \geqslant d)$		
R_3	$z \leftarrow z - d, \; d \leftarrow d + q_{(1)}$ $\cdots\cdots\cdots\cdots\cdots\cdots$	δ	
R_4	$d \leftarrow d_{+1}$ $\cdots\cdots\cdots\cdots\cdots\cdots\cdots\cdots\cdots\cdots\cdots$ $q_{(2\tau-1)}$	$\cdots$ $R_2 \wedge q_{(2\tau-1)}$	

2.7.4 Design layout

Three tables must be produced for a complete detailing of the unit.

(i)   Table 2.7.2 is a glossary giving the notation used for each register, counter and bistable, a list of the facilities provided and the points of connection required. Reference to Chapter 3 may be necessary before examining the table in detail.

(ii) Table 2.7.3 is the complete program table showing on the left-hand side the main program and all the subprograms. On the right-hand side each of the connecting points defined in Table 2.7.2 is given a separate column. For each row (program step), the entry in the column gives the source of the input to the connecting point demanded by the computation in the program step. Where a bistable is concerned, the upper entry in the square indicates the condition for its setting, the lower for its resetting.

(iii) Table 2.7.4, is derived from Table 2.7.3 column by column. For each connecting point, the complete source is the disjunction of all the entries in the column gated with the program step of the row in which the entry appears.

Actual circuitry will be given in 3.4, only for the multiplication and BCD - Binary Conversion subprograms. It suffices to say that this detailing is very much a routine translation into actual hardware. Once the registers, counter, bistables and multiplexers have been laid out, together with the facilities for operation on them defined in Table 2.7.2, it remains only to give the gating arrangements for Table 2.7.4. It is not necessary to show interconnecting links, since everything has been uniquely defined symbolically.

TABLE 2.7.2 GLOSSARY OF REGISTERS, COUNTERS, MULTIPLEXERS, BISTABLES, SIGNALS

Registers

Name	Type	Resolution	Add and clear (0)	Add or subtract (1)	Reverse diminuend and subtrahend (2)	Access point at -2 (3)	Double shift (4)	Suppress shift (5)	Change from double length to single length (6)	Reset (R)	Serial clear (C)	Arithmetic input (I)	0 Add 1 Subtract (Γ)	Shift suppress (S)	Double shift (D)	Main program (M)	BCD - Binary Conversion (P)	Multiplication (Q)	Square root (R)	Binary - BCD Conversion (S)
X	Shift reg.	20	1	1	1					1		1	1			1				
Y	"	20	1	1	1					1		1	1			1				
B	"	20	1			1		1			1	1		1		1	1	1		
C	"	20	1			1	1	1			1	1		1	1	1	1	1		
D	"	40(M14) 20(M14)	1	1				1	1	1		1	1		1	1	1	1	1	1
Z	"	40	1	1							1	1	1			1		1	1	

Counters / Multiplexers

Name	Type	Resolution	Modulo (20) reset (0)	Modulo (40) reset (1)	Down count (2)	Complete decode (3)	Double count (4)	Reset (R)	Preset (P)	Up count (U)	Down count (D)	Additional count (C)	Count enable (E)	Main program (M)	BCD - Binary Conversion (P)	Multiplication (Q)	Square root (R)	Binary - BCD Conversion (S)
T	Up count	6	1	1				1		1				1	1	1	1	1
Q	"	6		1		1		1	1	1		1				1	1	
I	Down count	3			1	1			1		1		1					1
A	Up count	6 × 4						1		1			1					1
U	6 × 2 + 3 Multiplexer															1		
V	"															1		

Bistables / Signals

Name	Type	Resolution	Clocked (0)	Comparator set, reset (1)	Preset, pre-reset (2)	Set, reset (R)	Preset, pre-reset (P)	Main program (M)	BCD - Binary Conversion (P)	Multiplication (Q)	Square root (R)	Binary - BCD Conversion (S)
β	Clocked flip flop		1	1	1	1	1					1
ξ	"		1	1		1			1			
η	"		1	1		1			1			
ψ	Bistable					1					1	
δ	Clocked flip flop		1	1	1	1	1				1	1
α	"		1			1		1	1	1	1	1
ST	Data Strobe							1				

Table 2.7.4 Schedule of source gating.

#	Symbol	Expression	Expression 2
1	X_R	MO	
2	X_I	$b_o \wedge$ (M2 $\vee$ M6)	
3	X_Γ	$\overline{X_\Gamma}$ MO	X_Γ M6
4	Y_R	MO	
5	Y_I	$c_o \wedge$ (M2 $\vee$ M6)	
6	Y_Γ	$\overline{Y_\Gamma}$ MO	Y_Γ M6
7	B_C	M7 $\vee$ M10 $\vee$ PO	
8	B_I	$(x_o \wedge$ M7) $\vee$ $(y_o \wedge$ M10) $\vee$ $(b_{-2} \wedge$ P1) $\vee$ $(u^i \wedge$ P2)	
9	B_S	$t_{(o)} \wedge$ P1	
10	C_C	M7 $\vee$ M10 $\vee$ PO	
11	C_I	$(x_o \wedge$ M7) $\vee$ $(y_o \wedge$ M10) $\vee$ $(c_{-2} \wedge$ P1) $\vee$ $(v^i \wedge$ P2)	
12	C_S	$t_{(o)} \wedge$ P1	
13	C_D	$q_{(o)o} \wedge$ QO	
14	D_R	M7 $\vee$ M10 $\vee$ RO	
15	D_I	$(c_o \wedge \psi \wedge$ QO) $\vee$ $(q_{(39)} \wedge$ R1) $\vee$ $(q_{(1)} \wedge$ R3) $\vee$ $(\zeta \wedge$ S2)	
16	D_Γ	$\overline{D_\Gamma}$ MO $\vee$ R3	D_Γ R1 $\vee$ S2
17	D_D	$t_{(o)o} \wedge$ (R2 $\vee$ R4)	
18	Z_C	M9	
19	Z_I	M9 $\vee$ M12 $\vee$ R3	
20	Z_Γ	$\overline{Z_\Gamma}$ MO	Z_Γ R3
21	Q_R	MO $\vee$ RO	
22	Q_P	$q_{(39)} \wedge$ QO	
23	Q_U	QO	
24	Q_C	$(\overline{q}_{(o)o} \wedge$ QO) $\vee$ $[(t_{(o)o} \vee t_{(1)o}) \wedge$ R2]	
25	I_P	$(6 \wedge$ PO) $\vee$ $(5 \wedge$ SO)	
26	I_D	P1 $\vee$ S3	
27	A_R	SO	
28	A_U	S2	
29	U	P2	
30	V	P2	
31	β	$\overline{\beta}: (d_o \leftrightarrow \zeta_o) \wedge$ S1	$\beta: (d_o \leftrightarrow \zeta_o) \wedge$ S1
32	β_P	$\overline{\beta_P}$	$\beta_P: t_{(o)o} \wedge$ S1
33	ξ	$\overline{\xi}: (x_o \leftrightarrow b_o) \wedge$ M5	$\xi: (x_o \leftrightarrow b_o) \wedge$ M5
34	η	$\overline{\eta}: (y_o \leftrightarrow c_o) \wedge$ M5	$\eta: (y_o \leftrightarrow c_o) \wedge$ M5
35	ψ	$\overline{\psi}: q_{(o)} \wedge$ QO	$\psi: (b_o)_{q(20)} \wedge$ QO
36	δ	$\overline{\delta}: (z_o \leftrightarrow d_o) \wedge$ R2	$\delta: [(z_o \leftrightarrow d_o) \wedge$ R2] $\vee$ $(d_o \wedge$ S3)
37	δ_P	$\overline{\delta_P}: t_{(o)o} \wedge$ S3	$\delta_P: t_{(o)o} \wedge$ R2
38	α	$\overline{\alpha}: (q_{(39)} \wedge$ P2) $\vee$ $(i_{(o)} \wedge$ P2) $\vee$ $(\overline{\delta} \wedge$ S3)	$\alpha: \{[(MO \vee M3) \wedge$ ST] $\vee$ M7 $\vee$ M10 $\vee$ M12$\} \wedge$ ECP
39	A_E	S2	

Table 2.7.3 Complete program   pp. 67—70

Main program	Sub program	ECP	Function	No of CP	α*	ISE	OSE		
M0		$t_{(19)}$	Idle	1	0	START	M14	SLEEP	M0
M1	P	$t_{(19)}$	BCD − Binary Conversion Input	8	1	ST			M1
M2		$t_{(19)}$	$x \leftarrow b,\ y \leftarrow c$	1	0	$\bar{\alpha}$			M2
M3		$t_{(19)}$	Idle	1	0	1		SLEEP	M3
M4	P	$t_{(19)}$	BCD − Binary Conversion Input	8	1	ST			M4
M5		$t_{(19)}$	$\xi \leftarrow (x < b),\ \eta \leftarrow (y < c)$	1	0	$\bar{\alpha}$		$x \leftarrow b\ y \leftarrow c$	M5
M6		$t_{(19)}$	$x \leftarrow [(x - b) \wedge \bar{\xi} \vee (b - x) \wedge \xi],\ y \leftarrow [(y - c) \wedge \bar{\eta} \vee (c - y) \wedge \eta]$	1	0	1			M6
M7		$t_{(19)}$	$d \leftarrow 0,\ b \leftarrow c \leftarrow x$	1	0	1		ECP	M7
M8	Q	$t_{(19)}$	Multiply $d \leftarrow x^2$	20	1	1			M8
M9		$t_{(39)}$	$z \leftarrow d$	2	0	$\bar{\alpha}$			M9
M10		$t_{(19)}$	$d \leftarrow 0,\ b \leftarrow c \leftarrow y$	1	0	1		ECP	M10
M11	Q	$t_{(19)}$	Multiply $d \leftarrow x^2$	20	1	1			M11
M12		$t_{(39)}$	$z \leftarrow z + d$	2	0	$\bar{\alpha}$		ECP	M12
M13	R	$t_{(39)}$	Square root $d \leftarrow \sqrt{z}$	40	1	1			M13
M14	S	$t_{(19)}$	Binary − BCD Conversion of $d(20)$ into A	63	1	R4			M14
				170					

P0		$t_{(19)}$	$b \leftarrow c \leftarrow 0,\ i \leftarrow \mu$			M0 ∨ M3			P0
P1	BCD − Binary	$t_{(19)}$	$b \leftarrow (b_{-2} + b)_{-1},\ c \leftarrow (c_{-2} + c)_{-1},\ i \leftarrow i - 1$			P2 ∧ $\bar{i}_{(0)}$			P1
P2		$t_{(19)}$	$b \leftarrow b + u^i,\ c \leftarrow c + v^i$						P2
P3		$t_{(19)}$	Exit			$i_{(0)}$			P3

Q0	Multiplication	$t_{(19)}$	$(b \wedge \varepsilon_t)q(20) \rightarrow [d \leftarrow d \wedge \varepsilon_t + c \wedge \varepsilon_\tau \mid q]q(20), q(21)\ldots q(39)$ $(q \leftarrow q + 1)q(0)$			M7 ∨ M10			Q0
Q1		$t_{(19)}$	Exit			$q_{(39)}$			Q1

R0		$t_{(39)}$	$d \leftarrow 0,\ q \leftarrow 0$			M12			R0
R1	Square root	$t_{(39)}$	$d \leftarrow d - q_{(39)}$			$[(R2 \wedge \delta) \vee R3] \wedge \bar{q}_{(39)}$			R1
R2		$t_{(39)}$	$(d \leftarrow d_{+1})t_{(0)},\ (q \leftarrow q + 1)t_{(0)}, t_{(1)},\ \delta \leftarrow (z > d)$			δ			R2
R3		$t_{(39)}$	$z \leftarrow z - d,\ d \leftarrow d + q_{(1)}$			$q_{(39)}$	$R2 \wedge q_{(39)}$		R3
R4		$t_{(39)}$	$d \leftarrow d_{+1}$, Exit			$q_{(39)}$			R4

S0		$t_{(19)}$	$A \leftarrow 0,\ i \leftarrow \mu - 1$			R4			S0
S1	Binary − BCD	$t_{(19)}$	$\beta \leftarrow (d > \zeta)$			$S2 \vee (S3 \wedge \delta)$			S1
S2		$t_{(19)}$	$d \leftarrow d - \zeta,\ a^i \leftarrow a^i + 1$			β			S2
S3		$t_{(19)}$	$i \leftarrow i - 1,\ \delta \leftarrow \overline{(d = 0)}$			$S1 \wedge \bar{\beta}$			S3
S4		$t_{(19)}$	Exit			$\bar{\delta}$			S4

2.8   ADDITIONAL MATERIAL ON PROGRAM ORGANISATION AND SYNCHRONIZATION

The examples which have been worked out in detail have been computational ones. Comments on the control of mechanisms have had to be more
general. It would be easy to take an oversimplified view of the program control of mechanical operations. A simple program control might require mechanical devices which are difficult to realise in practice. All too often
the mechanisms have already been designed and their characteristics have to
be accommodated in the control system, at the cost of complexity in the
latter.

For example, it may be desirable from the point of view of digital control to switch certain drives off and on according to program steps. In
practice, such switching of mechanical actions is frequently undesirable.
Even electromagnetic clutches are not always satisfactory. The ideal of
time independence of the sequence of operations cannot always be achieved.

2.8.1   Autonomous actions and sequences of actions

The subdivision of a control operation into program steps is princi-
pally governed by Out of Sequence Entries. To associate a program step with
every action would be wrong in principle. It has been shown in 1.3 that a
complete chain of actions can proceed autonomously under asynchronous control provided the sequence is unique. Such a sequence can be incorporated
in a synchronous program controller. The program controller starts the
sequence at the beginning of a program step and only proceeds beyond it when
the autonomous sequence has ended.

In some cases, operations or sequences can even run autonomously during
a number of program steps. Quite separately, a condition created by the
sequence is examined by the program control and an Out of Sequence Entry
obeyed. For example, components might be accumulated at some point and
counted in an autonomous counter. The program control is only concerned with
the condition that the required number has been accumulated. Only the counting of whole batches is registered by a counter which depends on program
control.

In fact, here the counting takes place not during a program step, but
during the transition from one step to another. The termination signal itself is used for such a once-only registration.

The program table (2.7.3) would in the general case have three additional columns. One, adjacent to the function entry, is used to give

autonomous and ancillary actions. Two further columns, adjacent to those giving ISE and OSE list actions at the ISE and OSE respectively. Where applicable,the same actions appear in both the latter columns.

2.8.2 Read only memories (ROM)

A facility which gives a significant degree of flexibility in the control system is the Read Only Memory. As an example, a control system might want to handle a number of different assemblies. The particulars of any one assembly can be coded in a Read Only Memory and the control system guided by the contents of that memory. The details of the coding depend upon the application.

It is convenient to follow the practice of computer organisation in alternating program steps which read the memory,with those that act upon what has been read. Generally, memories have addressing facilities and are organised in selectable words of 4 or 8 (sometimes single) bits. A read program step would bring down one or more words at a time, decode them and accordingly set stores for the control of the executive step.

Addressing the memory can be progressive, with each program step calling down a given number of successive words. Within the step, at the end of each clock period, the address is advanced. The step terminates when the required number of words has been read. To route the words to their appropriate destination a decoded counter is used. It is zeroed on entry into the program step and counted as the address is advanced. Clearly OSEs mean resetting the memory address,and program steps which do not use the memory leave the address unchanged.

More generally, the number of words to be read by a program step is allowed to vary. This means that the memory itself must contain addressing information. One method is to insert say a zero word into the memory, at the end of the string of words to be read by any program step. In this case, zero content becomes the terminator for the steps.

Whenever a program step is entered,to which there will be a backward OSE, the memory address is set into a temporary store (action under ISE). When the OSE occurs, the memory address is reset from the temporary store (action under OSE). For a forward OSE the memory address for the program step entered must be given in the memory.

In the example of assembly control, the Read Only Memory might contain the data for all the assemblies. They would appear as separate blocks of words (generally of variable length). It follows, that the initial address of each block has to be obtained from the memory, before the choice of block is made. Seldom, if ever, is there a case for searching a Read Only Memory. It is a fixed store, and all the necessary directions for accessing its content,should be contained within it.

2.8.3 Synchronization of independent program controllers

The normal method of synchronizing separate program controllers is to have a common clock. A different solution has to be adopted where it is undesirable to depend on a common clock. An example is a master controller with a number of slave controllers,where the latter are expected to be able to run autonomously. The problem is how to assure the synchronization of the separate clocks when the slave is brought under control of the master. The technique is to lock the clocks in what is called a phase-locked loop*.

In 4.3.2 phase-locked loops are introduced as hardware elements in AC signal processing. Here only the principle of its operation needs to be explained.

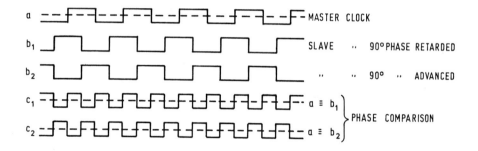

Fig 2.8.1 Synchronization by phase-locked loop.

In the timing diagram of Fig 2.8.1, line 'a' shows the clock used by the controllers (in the diagram the master clock). Its waveform is a square wave of unit amplitude. The mean level is shown hyphenated. If the clock of a slave controller (line b_1 or b_2) is in phase quadrature, equivalence ($\equiv$) of the two waveforms gives a clock of twice the frequency (line c_1 or c_2 respectively). For both cases the mean level (shown hyphenated) is the same in the original clock. In a phase-locked loop the frequency of an oscillator is proportional to this mean level. Such an oscillator provides

---

*In the form used here it is also called a correlation loop.

a clock which can thus be run either independently or in synchronism with another clock of the same frequency. The quadrature phase displacement of the two clocks is not important, since there is a one to one correspondence between clock periods, and they never overlap.

In practice each controller is run from a phase-locked loop oscillator with the phase comparison unconnected (equivalent to a high level input). Whenever a slave controller is to be brought under master control, the clock of the latter is taken to the phase comparison. The slave clock will settle into synchronism. Any drift from this condition will raise or lower the mean level of the phase comparison output and adjust the frequency of the slave controller,until synchronism is restored.

3   LOGIC CIRCUITRY

It is impossible to do the actual hardware design without consulting
the manufacturers' catalogues. The available devices do not follow a uniform
pattern, and unexpected difficulties can be encountered in their use. These
can only be anticipated by a careful study of data sheets.

The hardware description in this chapter will therefore be somewhat
simplified, as the main purpose of the chapter is to show how the techniques
of program control are realised in hardware.

3.1   BASIC ELEMENTS

The basic elements of logic hardware, the gates, perform Boolean
operations on the signals applied to their inputs. Sixteen Boolean operations
were listed in 2.2.1, but it was pointed out that only two of these are
fundamental.

Most modern logic series provide the operators AND, NAND, OR, NOR, as
well as XOR ($\neq$), and NXOR ($\equiv$). The symbols used in the technical literature
to denote these gates are given in Table 3.1.1.

Table 3.1.1 Symbols used to represent logic gates.

Inverter	AND	NAND	OR	NOR	XOR (Exclusive OR)	NXOR (Exclusive NOR)

It is arguable whether the use of the separate symbols given in
Table 3.1.1 is really helpful, particularly in an explanatory text. The
alternative is to chose two fundamental operators and simply represent these
as square blocks.

In the present text the square blocks denote NAND gates and, in the
case of a single input, Inverters. Their function as ANDing or ORing gates
is assumed obvious from context. It is made easy by the naming of signals.

76

Whenever the name is quoted, the signal is assumed to be true, the name barred indicating the reverse. Thus if a 2-input NAND gate has, say, signals P1 and P2 on its inputs, its function must be ANDing, with the output $\overline{P1 \wedge P2}$. If the inputs are $\overline{P1}$ and $\overline{P2}$, its function must be ORing, with the output $P1 \vee P2$.

3.1.1 Gates

The basic gating element is the one-or-more-input NAND gate (Fig 3.1.1).

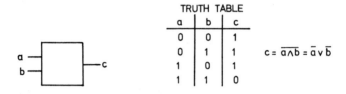

TRUTH TABLE

a	b	c
0	0	1
0	1	1
1	0	1
1	1	0

$c = \overline{a \wedge b} = \bar{a} \vee \bar{b}$

Fig 3.1.1 NAND for 1, ORNOT for 0, OPEN INPUTS = 1.

The single-input NAND gate is used in the same form to represent the inverter. The distinction between input and output is assumed obvious from context.

Wired-OR is used extensively because, apart from economising on components, it simplifies the diagrams. It is shown by the simple connection of outputs as in Fig 3.1.2, where the equivalent is also given.

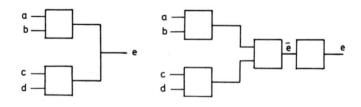

Fig 3.1.2 Wired-OR. $e = \overline{(a \wedge b) \vee (c \wedge d)}$ and equivalent.

3.1.2 Flip Flops

Flip Flops can be either clocked or unclocked, single-stage or 2-stage (master-slave).

A single-stage unclocked version is the Eccles-Jordan bistable* formed by cross-coupling 2 NAND gates as shown in Fig 3.1.3.

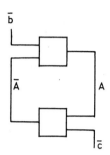

Fig 3.1.3 Eccles-Jordan bistable or NAND R/S latch.

As long as b and c are not simultaneously 1, A is set by b, $\bar{A}$ is set (A reset) by c.

A form of bistable which is intermediate between the single-stage unclocked and 2-stage clocked Flip Flop is the latching bistable (sometimes called D-type Flip Flop) of Fig 3.1.4.

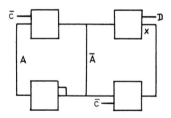

Fig 3.1.4 Latching bistable or D-type Flip Flop.

Two NAND latches have one side commoned in a wired-OR.

Initially c = 1. Since c→x, A ≡ D. There is no bistable action. As soon as c changes to 0, bistable action takes over. Moreover, the left-

---

*A more accurate name would be NAND R/S latch. The abbreviation R/S stands for Reset/Set.

hand bistable dominates. Since $x \equiv A$, D has no effect when $A = 0$. If $\bar{A} = 0$, it is held in this state by the left-hand bistable and D is unable to change it. The result is that the last state of D before the transition of c from 1 to 0 is recorded in A.

The most common 2-stage (master-slave) clocked Flip Flop is known as the JK Flip Flop. Fig 3.1.5 shows its composition.

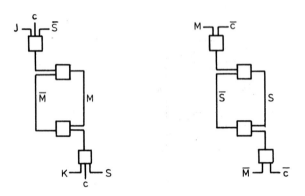

Fig 3.1.5 Composition of JK Flip Flop.

When $c = 1$, the master (M) is set according to the inputs J, K and the previous state of S. When $c = 0$, the slave (S) takes the state of M. Denoting no change from the previous state by n.c., the result is given in Table 3.1.2.

Table 3.1.2 Definition of JK Flip Flop.

Present State of Flip Flop		$\bar{S}$	S
State of Inputs	State of Clock	New State of Flip Flop	
$\bar{J}$, $\bar{K}$	c	n.c.	n.c.
	$\bar{c}$	$\bar{S}$	S
J , $\bar{K}$	c	M	n.c.
	$\bar{c}$	S	S
$\bar{J}$, K	c	n.c.	$\bar{M}$
	$\bar{c}$	$\bar{S}$	$\bar{S}$
J , K	c	M	$\bar{M}$
	$\bar{c}$	S	$\bar{S}$

The JK Flip Flop is also provided with an asynchronous Set and Reset which act on both M and S and hence override the above setting. The diagrammatic representation of Fig 3.1.6 will be used henceforth.

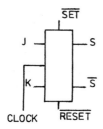

Fig 3.1.6 Diagrammatic representation of JK Flip Flop.

The subject of synchronous Flip Flops has been deliberately simplified in the diagrammatic representation of Figs 3.1.4 and 3.1.5.

It will be observed that in a JK Flip Flop the non-equivalence of inputs, J $\neq$ K, reduces it to a D-type Flip Flop so that after transition of c from 1 to 0, S $\equiv$ J. Thus only D-type operation need be considered in describing the following 4 variants.

(i)   Single stage Flip Flop - clocked
(ii) Single stage Flip Flop - edge triggered
(iii) Master slave operation - clocked
(iv) Master slave operation - edge triggered.

i and ii are generally (though not always) referred to as latches. Prior to the transition of c from 1 to 0, the output follows the state of the input. After the transition it becomes independent of the input.

iii and iv take their output from the state of the master at the time of the transition of c from 1 to 0. Prior to this the output is independent of the input and holds its previous setting.

i and iii set a bistable irreversibly if the input goes high while c = 1. Thus the occurrence of a high input during c will determine the output during $\bar{c}$.

ii and iv do not set a bistable until the transition of c from 1 to 0 actually takes place. The output is determined by the input state at the instant of the transition. There are many different circuits which give the above properties. The details do not matter except where it may be necessary

to refer to them in order to make quite sure to which particular class a
Flip Flop belongs. There are also minor variants such as the polarity of
inputs and clock. These do not affect the above classification.

3.1.3 Clock and characteristic period generator

There are various ways of generating master clock pulses. Three
determining factors are quoted below and the appropriate clock mechanism is
given in each case. Further details will be found in the references.

(i)   An exact frequency is important: Crystal Oscillator
(ii)   There is a need for synchronising with other clocks (see 2.8.3):
       Phase Locked Loop [3.1]
(iii) Mark-Space ratio of clock is to be other than unity:
       Astable Multivibrator [3.2]

The first stage beyond this point is generally the generation of a
2-phase clock by the circuitry of Fig 3.1.7.

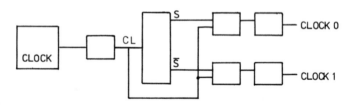

Fig 3.1.7 Generation of 2-phase clock

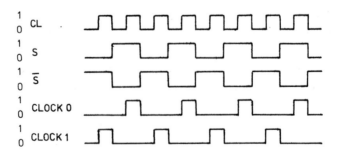

Fig 3.1.8 Timing diagram for Fig 3.1.7

The frequency is halved and two non-overlapping pulse trains CLOCK 0 and CLOCK 1 are generated. A timing diagram for circuit of Fig 3.1.7 is given in Fig 3.1.8.

Where a 4-phase clock is needed, the circuit of Fig 3.1.7 is extended to that of Fig 3.1.9.

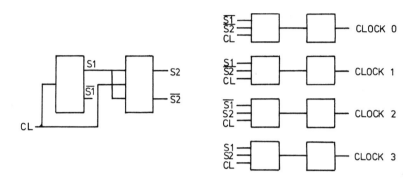

Fig 3.1.9 Generation of 4-phase clock.

The decoder on the right hand side of Fig 3.1.9 would normally take the form of an integrated 1 out of 4 decoder (for example type 9321) with the clock (CL) tied to the Enable.

The generation of the characteristic period (CP) requires a modulo (n) counter for a programme which uses n component vectors. As an example, if n = 20 the circuit of Fig 3.1.10 may be used, although in practice integrated binary counters (for example type 9316) would be preferred.

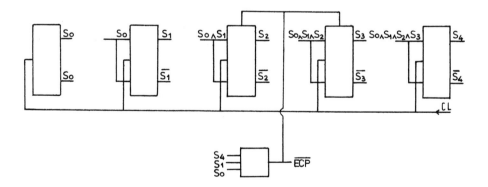

Fig 3.1.10 Generation of characteristic period for resolution 20.

The 5 Flip Flops,whose J and K inputs are successively the conjunction of the outputs of the preceding stages,form a modulo (32) synchronous counter. However when $S4 \wedge S1 \wedge S0 = 1$, which means the count has reached 19, an End of Characteristic Period signal (ECP) is formed. This sets the remaining stages which are not set to 1 (S2 and S3). The ECP signal is unaffected and remains set for the current clock period. At the end of this period the counter returns to 0. The arrangement has converted a modulo (32) counter into a modulo (20) counter.

3.2   PROGRAM CONTROL

Sufficient circuitry has been developed in the last section to explain the techniques of program control.

3.2.1   Program steps

The interconnection of JK Flip Flops shown in Fig 3.2.1 called a Ring Counter is the basic form of program control.

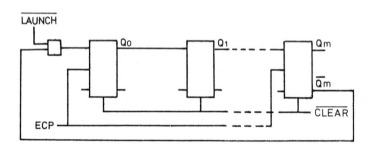

Fig 3.2.1   Ring counter with launch.

If all Flip Flop stages are initially reset $(\overline{Q_0}, \overline{Q_1} \ldots \overline{Q_m})$, a signal LAUNCH, appearing once while ECP is present, launches the counter. At the beginning of next characteristic period the state is $Q_0, \overline{Q_1}, \ldots \overline{Q_m}$.

After the end of this period the state is $\overline{Q_0}, Q_1, \overline{Q_2} \ldots \overline{Q_m}$ and so on until after another m periods the state is again $Q_0, \overline{Q_1}, \ldots \overline{Q_m}$.

If $Q_0, Q_1 \ldots Q_m$ are successive program steps, the circuit controls the program for the case where all program steps are internally controlled, their duration is one Characteristic Period and there are no OSEs.

3.2.2   Terminations

The transfer between program steps Q_i and Q_{i+1} as a result of the termination q_i is given in Fig 3.2.2. Two alternative forms are shown. (b) is the preferred form which will be used below, except where (a) is clearer from a descriptive point of view.

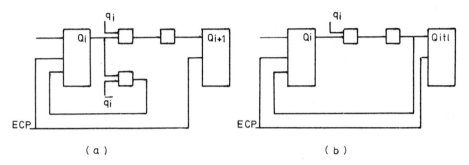

Fig 3.2.2   Termination of program step Q_i: 2 methods.

The state Q_i, $\overline{Q_{i+1}}$ is held while $q_i = 0$. The first change of q_i during an ECP resets Q_i and sets Q_{i+1}.

An important feature of the arrangement is that q_i need not be a logic signal. It could be the output of a microswitch or other sensor connected directly into the logic circuitry*. Q_i enables some external action, q_i signals its termination. Once the termination is accepted (during the first ECP which coincides with it) the subsequent state of the switching signal is of no consequence**.

There is a caveat. Supposing a random signal q_i occurs just before the transition of ECP to $\overline{\text{ECP}}$. The specification of a JK Flip Flop stipulates that a signal on J or K will be recognised if it is a present for a minimum period before the clock transition (in the present case of ECP). This minimum varies from 10-100 ns depending on the technology used in the components. What is more important here, is that the minimum time might be very slightly longer in the case of Q_{i+1} than Q_i. If q_i occurs between the instants determined by this variation, it will be recognised for the

---

*A buffer relay or other element (see 4.2.2) is generally necessary if the sensor is remote.
**This avoids the use of retriggerable monostables which are normally used to convert the switching signal into a one shot logic signal. Monostables are particularly sensitive to transient noise and should, where possible, be avoided.

resetting of Q_i, but not for the setting of Q_{i+1}. This results in a state $\overline{Q_i}, \overline{Q_{i+1}}$ from which the program controller cannot recover.

The simplest method of overcoming the difficulty is to rely on synchronisation with a 2-phase clock. If the characteristic period generator is clocked by CLOCK1, q_i can be gated with CLOCK0. The output of a bistable (Fig 3.1.3) set by $q_i \wedge$ CLOCK0 (and reset together with Q_i) should take the place of the signal q_i in Fig 3.2.2.

A more obvious pitfall is the case of a cyclic program control in which all program steps other than Q_i are computational and of short duration. q_i must obviously have returned permanently to 0 by the time the cycle has returned to Q_i. This difficulty does not arise if at least two program steps are dependent on external terminations.

One big advantage of the clocked program controller method of interrogating sensor signals results from using a clock (ECP) whose mark space ratio is very much less than unity. In a noisy environment a common difficulty is the appearance of spurious signals on sensor lines. If these spurious signals have an amplitude comparable with that of the genuine signal, there is no way in which the logic can distinguish between them. However they are usually of very short duration. So one must rely on reducing the probability of detecting spurious signals. This results from driving the program controller with very short clock pulses (for example, 100-500 ns) and leaving the longest possible spaces during which spurious signals can have no effect.

Summarising, Q_i may be a program step which controls

(i)   an external action terminated by a sensor signal q_i

(ii) an internal operation of one characteristic period duration terminated unconditionally by ECP

(iii) an internal operation of more than one characteristic period duration terminated by some internal logic condition.

In each case the program step might have to be subjected to internal or external enabling (see 2.5, also 3.2.6-8). In case i the enabling of the termination follows from the enabling of the external action. In cases ii and iii the termination must itself be enabled.

It will be sufficient to illustrate the external enabling of case ii (Fig 3.2.3) to draw attention to the synchronisation of the enabling signal.

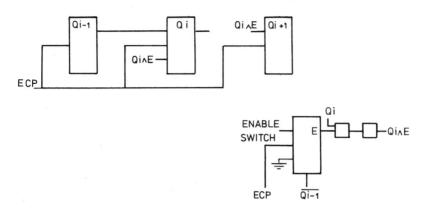

Fig 3.2.3 Enabling a computational step.

The important point to note in this instance is that E must be clocked by the same ECP as the program controller. This assures that E is set at the beginning of a Characteristic Period. The Q_i computation is then enabled for a full period at the end of which the termination $Q_i \wedge E$ transfers control to Q_{i+1}.

3.2.3 OSEs

The simplest case of an OSE is the omission of one or more program steps on a condition (say α). In Table 3.2.1 the program is shown in the usual way and Fig 3.2.4 gives the circuitry.(Clock signals are omitted henceforth)

Table 3.2.1   Program for: Pl, If (α) then P2, P3; P4

P1	ISE	OSE
P2	α	
P3		
P4		$P1 \wedge \bar{\alpha}$

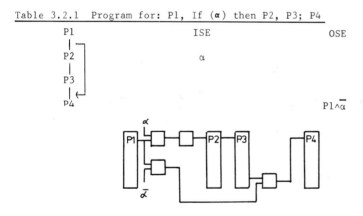

Fig 3.2.4   Circuitry for program of table 3.2.1 (Reset Signals omitted)

86

It should be noted that the gate between P3 and P4 functions as an OR gate. Inputs are $\overline{P3}$ and $\overline{P1_\wedge\overline{\alpha}}$. The output is $P3 \vee (P1_\wedge\overline{\alpha})$ as required by Table 3.2.1.

Slightly more complicated cases can be illustrated by giving the circuitry for two of the programs of Section 2.3. The ones chosen are Table 2.3.4, Fig 3.2.5 and Table 2.3.5, Fig 3.2.6.

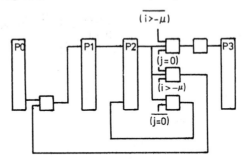

Fig 3.2.5 Circuitry for program of Table 2.3.4.

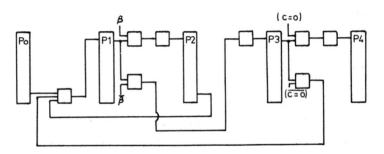

Fig 3.2.6 Circuitry for program of Table 2.3.5.

The If then ... Else ...; form of OSEs is illustrated in Fig 3.2.7 which gives the circuit for the hypothetical program of Table 2.4.4.

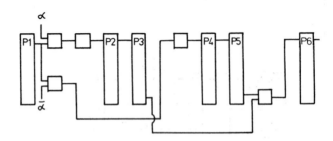

Fig 3.2.7 Circuitry for program of Table 2.4.4.

Finally two equivalent circuits are given in Fig 3.2.8 for the program of Table 2.4.5.

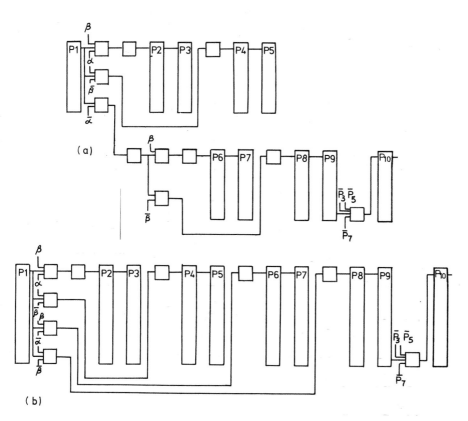

(a)

(b)

Fig 3.2.8   Two forms of the circuit for the program of Table 2.4.5.

3.2.4   Interlocks

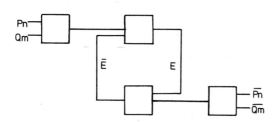

Fig 3.2.9   Setting of mutual enable of P_n and Q_m.

The circuit for the setting of the Mutual Enable $E = [E \wedge (Pn \vee Qm)] \vee (Pn \wedge Qm)$ is the Set-Reset bistable of Fig 3.2.9.

The more general form of more than two Mutual Enables and the looser coupling of overlapping sequences are obtained by the same mechanism.

The circuit for the inhibit setting (2.5.3) is again a Set-Reset bistable (Fig 3.2.10) but it functions differently in this case.

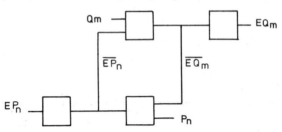

Fig 3.2.10 Mutual inhibition of P_n and Q_m.

The bistable is initially (before P_n and Q_m) in a 1,1 state. From this it falls into the EP_n, $\overline{EQ_m}$ state if P_n precedes Q_m; or the EQ_m, $\overline{EP_n}$ state if Q_m precedes P_n. It cannot reverse until respectively P_n or Q_m have disappeared. It returns to 1,1 state when both P_n and Q_m are completed.

Fig 3.2.11 gives the circuit for the case of an external linkage between two program steps from separate controllers. P and Q are brought together at R between P_{m-1} and P_m and Q_{n-1} and Q_n respectively. It is assumed that r, the termination of R, is a consequence of the action enabled by $E \wedge R$ (the timing diagram for the circuit was given in Fig 2.5.4).

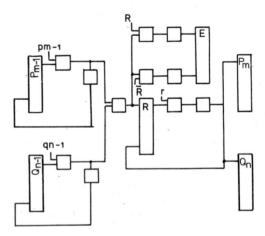

Fig 3.2.11 Commoning of program controllers at program steps (r is the termination of action enabled by $E \wedge R$).

3.2.5 Subprograms

The circuit of Fig 3.2.12 copes with the case of a main program whose Characteristic Period is an integral multiple of the subprogram period; which includes the case of equal periods (see 2.6.1 and Fig 2.6.1).

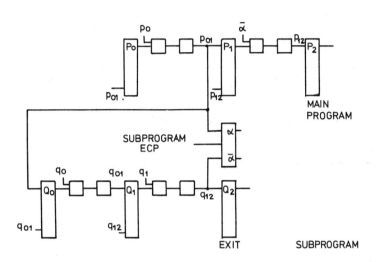

Fig 3.2.12 Subprogram control for the case of main program CP integral multiple of subprogram CP

The circuit which synchronizes the main program and subprogram registers in the case of unequal characteristic periods is given in 3.2.13. The corresponding timing diagram was given in Fig 2.6.2.

For a complete understanding of the circuit, reference should be made to Section 3.3, where Medium Scale Integrated devices are described. Contrary to the hardware described so far, both devices type 9316 and 9328 set their internal Flip Flops on the transition $\overline{\text{CLOCK}}$ to CLOCK. Hence 9316 is clocked by $\overline{\text{CLOCK}}$ and 9328 by $\overline{\text{CLOCK}} \vee \beta$. The latter is equivalent to $\overline{\text{CLOCK} \wedge \bar{\beta}}$, which agrees with Fig 2.6.2.*

---

*It is worth commenting here on the right method of deriving clock pulses which are subject to inhibition. The rule is to formulate the Boolean expression for the condition when the clock pulse is to be true, then invert it if the device works on the $\overline{\text{CLOCK}}$ to CLOCK transition.

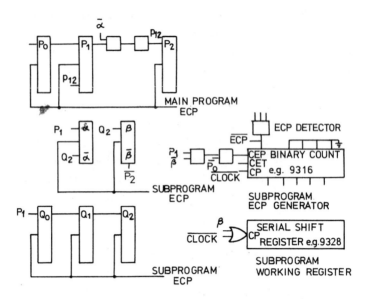

Fig 3.2.13 Subprogram control for the case of main program CP and subprogram CP unequal.

3.2.6 Start/Stop

In most control systems there is a need for a start/stop facility which permits a suspension of operation during normal cycling. A quite different procedure (3.2.7) applies to starting the system after initial switch on.

The point at which a control system cycle may be suspended is arbitrary since ideally, the control mechanism is time—independent. All program controllers must eventually come to rest because of interlocks, if the rules of Section 2.5 are adhered to. It is however best to choose a unique point for the suspension and the obvious one that suggests itself is the Mutual Enable of all the program controllers, such as P_n, Q_m, R_1 in Fig 2.5.2.

The timing of the stop signal is not critical, since the program controllers determine the exact time of suspension. The timing of the start signal is important since the release from suspension must always take place at the start of the characteristic period.

A STOP bistable is set by a stop signal and reset by a start signal. The stop and start switches are merely used to create the necessary pulses, the latter being synchronised by ECP to coincide with the start of CP.

Both sides of the alternate action switches are tied to earth so that pulses are created during the transition of the switch action.

The STOP bistable is also set by the CLEAR signal as part of initial setting. Thus the control system is started initially from the same condition that it is suspended in by a stop signal.

If the point of suspension of the program controllers is P_n, Q_m, R_1, $\overline{STOP}$ is applied to the setting gate of the Mutual Enable $Ep_nQ_mR_1$.

Fig 3.2.14 also shows ancillary equipment which is usually required. A control panel indicator driven from $\overline{STOP}$ indicates that the controllers are suspended in a condition from which they can be released by the start switch. Buffer relays are shown inserted, since the control panel is usually remote (see 4.2), +V indicates the supply voltage needed for their operation.

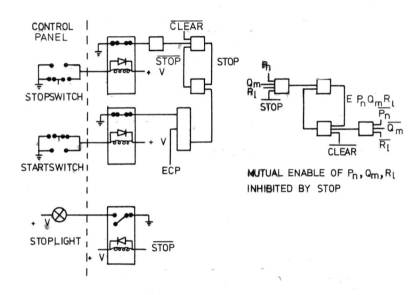

Fig 3.2.14 Start/Stop mechanism.

3.2.7 Clear and launch

When a control system is switched on, the transient conditions cause the memory elements to take up states which are completely indeterminate. Only the clock and characteristic period generator can be assumed to function correctly. To safeguard the system against unwanted actions and prepare it for operation the following principles should be adhered to.

(i) Actuators must not be powered until the controlling logic has first been correctly conditioned.

(ii) Since the state of all bistables is indeterminate when voltage is applied to the logic circuitry, at least one bistable (here called the ERROR bistable) must be conditioned by other than electrical means. This ERROR bistable can then be used to prevent the application of power to actuators.

(iii) A normally closed (NC) relay, connected to earth, can be used to bias the ERROR bistable. The relay is prevented from opening by the Off ($\overline{ON}$) state of the power supply switch. When the switch is moved to the ON state, the connection to earth through the relay will be broken. Due to the slow action of the relay, the connection will be opened some time after the proper voltage levels have been applied to the logic circuitry. After this the ERROR bistable is free to be reset.

(iv) A general CLEAR signal derived from the clear switch now sets all bistables into their correct initial state. Among these

    a the ERROR bistable is reset. This closes the normally open (NO) relay and provides power to the actuators

    b all program control Flip Flops are reset (see 3.2.1) and

    c the STOP bistable (see 3.2.6) is set to the STOP state.

(v) After the transients from the clear switch have disappeared a single LAUNCH pulse, coincident with ECP, sets the initial program step of all program controllers (see 3.2.1). If there are no intervening program steps which depend on external termination, the controller moves on to the next program step which is tied in a Mutual Enable.

(vi) The control system now suspends, because the Mutual Enable is gated with $\overline{STOP}$. From this it can be released by the START signal (see 3.2.6).

(vii) The start up cycle which usually precedes normal operation may need an additional 'once only' signal. A PRIME bistable serves for this purpose. Set by the LAUNCH pulse, PRIME gates the START signal which in turn resets the PRIME bistable.

The mechanism is essentially fail-safe if all fault conditions (including any emergency stop) set the ERROR bistable. Power is thereby removed from the actuators and this can only be restored by first issuing another CLEAR signal.

The three circuits used in this procedure are shown in Fig 3.2.15. The function of the ERROR bistable (Fig 3.2.15a) is self-explanatory from the above. The only other point to note is that there is no danger in the simultaneous powering of actuators and clearing of bistables. Since relay action (particularly in power relays needed for actuator power) is always slower than transistor circuitry, the logic will be conditioned before any actuator can react.

The circuit of Fig 3.2.15b generates a single LAUNCH pulse at the end of the transient from CLEARSWITCH. Its mode of operation is best explained with the help of the timing diagram of Fig 3.2.16. Here the signal from CLEARSWITCH (inverted) is deliberately shown accompanied by a bounce. This occurs after the ECP which follows the end of the main switch signal. The consequence is that the first LAUNCH pulse (which sets initial Program steps) is premature. However since the setting of the Program steps is immediately reversed before being set again, the program controllers cannot be wrongly launched.

The circuit of Fig 3.2.15c is the conventional circuit for a 'once only' pulse in which a bistable enables the pulse which resets it. Its function in the start up cycle is governed by the control system under consideration.

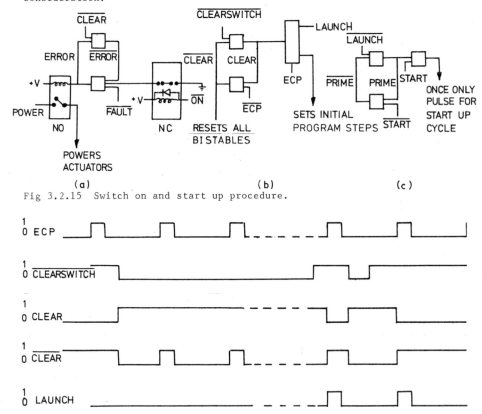

(a)　　　　　　　　　　　(b)　　　　　　　　(c)

Fig 3.2.15 Switch on and start up procedure.

Fig 3.2.16 Timing diagram for circuit of Fig 3.2.15b.

3.2.8 Single step

An essential aid to commissioning and maintaining a control system is a single step facility. It is particularly important for a decentralised controller in which a number of program controllers run simultaneously.

The single step or MANUAL mode of operation is obtained by

1   making the action governed by each program step dependent on an ENABLE,

2   making the transition between program steps reset the ENABLE, and

3   providing a SINGLE STEP SWITCH, which gives a pulse (SS) coincident with the beginning of the Characteristic Period, to set the ENABLE.

Since the program controllers are autonomous, a separate ENABLE Flip Flop must be associated with each program. Each of these Flip Flops must be accompanied by another Flip Flop called the INHIBIT. The latter is set when the program controller goes to the $\overline{\text{ENABLE}}$ state. In the INHIBIT state it awaits the SINGLE STEP signal for the transition to ENABLE. The two Flip Flops are necessary to synchronise the transitions to those of the program controller proper. In fact the best solution is for both to use the same mechanism based on the ECP signal (Fig 3.2.17).

Various additional features may be incorporated such as:

1   A mechanism by which transition from MANUAL to AUTO and vice versa may be made at any time during the running.

2   A distribution of SINGLE STEP pulses to the various program controllers dependent on switch settings,so that program controllers may be advanced individually as well as together.

3   An indicator for each program controller which lights when the appropriate controller is enabled.

4   Columns of Light Emitting Diodes one for each Flip Flop of the program controllers to indicate the operative program step.

5   A discriminatory single stepping so that during initial commissioning all program steps can be stepped and subsequently, for operator use, only those which have actuators associated with them.

Only the first of these features is included in the circuitry of Fig 3.2.17 as the remainder are matters of detail. The example concentrates on the correct method of enabling program steps, in particular computational steps and where there are OSEs. Special attention has to be paid to the problem of external linkages between mechanisms controlled by separate program steps (see end of 2.5.2). Separate single stepping of individual program controllers may not be possible in these cases unless the controllers are brought together to a common program step as suggested in 2.5.2.

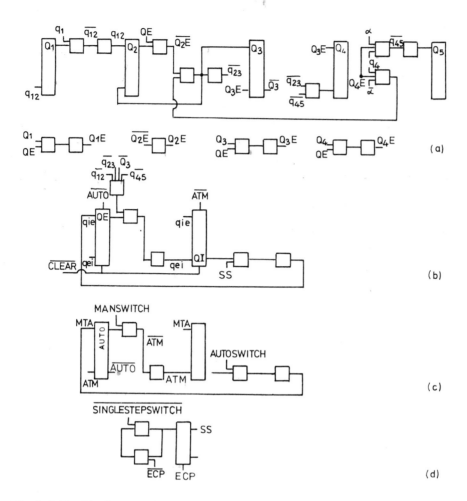

Fig 3.2.17 Single stepping mechanism for typical program controller.

Fig 3.2.17a is a typical program controller in which

1   the termination q_1 of Q_1 is dependent on action governed by Q_1E (Q_1 enabled),

2   Q_2 is a computational step of one Characteristic Period duration, but the transition to Q_3 has the alternative of an OSE from Q_4,

3   Q_3 is another computational step of one Characteristic Period duration, and

4   Q_4 is a computational step terminated by q_4 which branches to Q_5 or Q_3 depending on α .

Underneath the program controller the gates needed for enabling the program steps by QE are shown. QE is the enable for the stepping mechanism. Any other enables, such as interlock enables, would also appear on the gates. The active phases of the program steps are Q_1E, Q_2E, Q_3E. Where the termination is dependent on some external action, it can gate the program step itself (for example Q_1). Otherwise it must gate the enabled program step (for example Q_4).

Fig 3.2.17b shows the single stepping mechanism. During the Enable phase QE is set. Every one of the program step transitions resets QE and sets QI (Q Inhibit). The single step pulse SS derived from the circuit of Fig 3.2.17d reverses this to give a new Enable. The clock pulse on the QE and QI Flip Flops is ECP so that synchronism is maintained with the program controller. The Enable and Inhibit always become effective at the beginning of Characteristic Period. The initial CLEAR gives $\overline{QE}$, QI, so that either the AUTO or SINGLE STEP switches must be pressed before the START switch can have effect.

The circuit of Fig 3.2.17c processes changes in the setting of the MANUAL/AUTO switch. Again synchronism with the above is maintained. AUTO is applied to the asynchronous setting of QE (Fig 3.2.17b) so that it overrides other changes in the single stepping mechanism. A change to Manual (MAN) frees the single stepping mechanism so that it can respond in the manner described in the last paragraph. The transition signal AUTO to MAN (ATM) also resets QI, so that in the Manual state, QE and QI are never set simultaneously.

Fig 3.2.17d is a one shot circuit for the SINGLE STEP SWITCH and functions in the same way as the circuit of Fig 3.2.15b. There is a remote danger that the two pulses due to switch bounce illustrated in Fig 3.2.16 could step two program steps in succession. This could only happen if the first step was short and the second pulse arrived after its termination.

3.3   DERIVED ELEMENTS

Medium Scale Integration (MSI) devices are available to perform certain standard operations such as storing, counting, encoding, decoding. Their use reduces both the number of basic elements and interconnections needed in a control unit.

The composition of the MSI elements is dictated among other things by:
(i)   the maximum number of basic gating circuits which can be incorporated in a chip without exceeding permissible power dissipation in the package,

(ii) the number of external contacts available in the encapsulation (Dual-in-Line 14, 16 or 24 pin).

The former is significant in bipolar technology (DTL, TTL, etc), so that, for instance, the Shift Registers are limited to 16 stages (type 9328). The latter limits the number of stages in parallel registers and counters and the number of parallel lines in decoders and encoders.

In the following some typical applications of MSI devices are presented. The functions described are connected with the algorithms of Chapter 2. A much more extensive treatment will be found in [3.3].

3.3.1 Registers

Supposing a 22-bit register is required to hold a vector c and the algorithm, such as BCD-Binary Conversion (Section 2.3.2) calls for access to component c_{-2} as well as c_0 (Fig 2.3.2b). One way of satisfying the requirement is to use one Dual 8-bit Shift Register (type 9328) together with two Universal 4-bit Serial-Parallel registers (type 9300). The circulation would be arranged as in Fig 3.3.1a and the actual connections to the devices as in Fig 3.3.1b. To understand the latter, the data sheets [3.4] for the devices must be consulted.

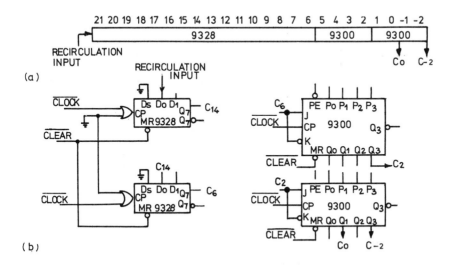

Fig 3.3.1 22-bit shift register C with additional access point C_{-2}.

3.3.2 Counters

The main form of counting is in binary progression. If a 4-bit counter terminates at 9, it serves as a decade counter. If it terminates at 15, it is a modulo (2^4) counter. Only the latter will be considered below.

Two methods of binary counting are used. Asynchronous, in which the carry from stage to stage is progressive; and synchronous, in which it is instantaneous. Asynchronous counters are rarely used, even though their speed is often quite adequate. This is because an interrogation phase is needed after the counting phase to safeguard against false counts being recognised in the transition state.

The synchronous counter is the most common. Taking first the form which only allows UP counting (type 9316), two important features should be observed.

(i) The Preset is synchronous. This means that the Parallel Enable (PE) is applied to the gating arrangement in front of the counter stages and overrides normal counting. The change of state of the counter stages (to that of the parallel inputs) takes place only at the normal clock transition.

(ii) The carry from the counter (TC), when the count has reached 15, is held irrespective of clock, but is gated with a Carry Enable (CET). If this is tied to the carry (TC) of the preceding counter, cascading of the devices to give a multistage synchronous counter of any required length is possible.

Fig 3.3.2 shows the use of two type 9316 devices to give a modulo (60) synchronous count, such as in a Characteristic Period generator for a double length 30-bit resolution computation. (A similar device, type 74163, has also a synchronous master reset (MR), to which $\overline{ECP}$ could be directly connected.)

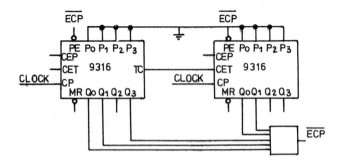

Fig 3.3.2 Modulo (60) synchronous count.

Two forms of 4-bit synchronous counters are made which allow both UP
and DOWN counting.

(i)   Type 9366 or 74193, in which separate Clock Inputs are provided for UP
counting (CPU) and DOWN counting (CPD). The carries again separate into UP
count carry (TCU) and DOWN count carry (CPD). They are gated with CPU and
CPD respectively and serve solely for cascading counters,in which case $\overline{\text{TCU}}$
is connected to CPU and $\overline{\text{TCD}}$ to CPD. There are not enough contact pins
available in a 4-bit UP/DOWN counter to provide separate Enables for the
clock input as in 9316 or 74163.

(ii)   Type 93191 or 74191, which has a single clock input and a steering
input (DN/UP) for the direction of counting. It has a carry output (MAX/MIN),
which is independent of clock. This is separately gated with clock and Count
Enable (E) and made available at an external contact as RIPPLE CLOCK. In
cascading,RIPPLE CLOCK is taken to the Enable Input (E) of the next device
if a common clock is used or to the clock input (CP) if a common Enable is
used.

The choice of device and method of use depends entirely on application.
Since the Preset and Reset in UP/DOWN counters are asynchronous, the method
of producing a modulo (n) counter is the one advocated in Fig 3.1.10. Care
must however be taken if a DOWN count follows the UP count termination or
vice versa, since the state of the counter would then not form part of the
legitimate sequence.

A set of synchronous counters with the same characteristics as the above
is available in a CMOS range $\left[3.5\right]$.

3.3.3 Multiplexers or encoders

Multiplexers have been referred to in Chapter 2, where they were used
to load parallel data into the serial computing unit.

Multiplexers are available in Dual 2-bit, 3-bit or 24-pin 4-bit configurations. A Strobe or Enable is provided. When the Strobe (S) is low it
blocks the connection between all the parallel inputs and the output. The
output is then low. But since otherwise the output is the same as the
selected input, it follows that it is impossible to distinguish non-selection
and selection of an input which is low. The output must therefore act in
only one state and this must be opposite to the non-selected state. The
active state must correspond to a high input.

Fig 3.3.3 shows 3 type 93151 devices used to transfer the content of
a 24-bit parallel store into a 24-bit Shift Register (consisting of $1\frac{1}{2}$ type

9328 devices). The loading of the serial register is governed by the LOAD
Flip Flop which is reset at the end of the Characteristic Period. The loaded
content is recirculated thereafter. An alternative to the scheme of Fig 3.3.3
is the use of a special device type 93165 which is an 8-bit Serial Output,
Serial or Parallel Input Shift Register.

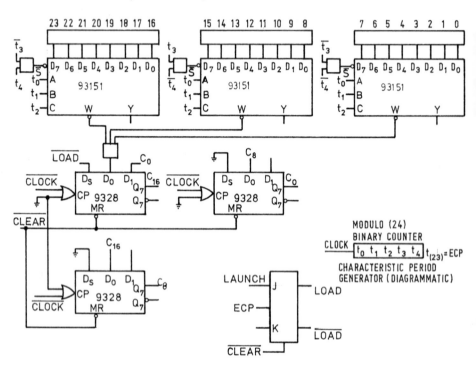

Fig 3.3.3 Parallel to serial conversion using multiplexer.

3.3.4 Demultiplexers or decoders

A large number of decoding devices is available to deal with the
various forms of codes used. They are dealt with fully in the literature.
It was shown in Fig 2.1.5 that a decoder also serves as a serial to
parallel converter, if the synchronising count is decoded and used to gate
the serial content of the shift register. However the same feature has to
be noted as with multiplexers. It is impossible to distinguish between non-
selected output and selected output corresponding to a low input. In this
connection there was a reference in Section 2.1 to the use of addressable or
decoding latches. Such a device (type 9334) is used in Fig 3.3.4 for serial
to parallel conversion.

The example of serial to parallel conversion of Fig 3.3.4 is the reverse of the scheme given in Fig 3.3.3. Again an alternative is the 8-bit serial input, parallel output shift register type 93164.

The multiplexers and demultiplexers described so far, select by gating the input and output respectively. Gating being unidirectional, the two devices are quite separate. The alternative is to use make/break switches. In that case, the one device will serve as both multiplexer and demultiplexer by simply interchanging input and output.

Switching multiplexers and demultiplexers are familiar in their mechanical form as uniselectors and commutators respectively. In more recent technology they occur as CMOS devices $[3.6]$. This facilitates their use in logic systems.

Non-selection means an open switch. But since a disconnected input to a logic device is equivalent to a 1, the active state of the switched signal must be a 0.

The thumbwheels used in Fig 3.4.2 of the next section are an example of bilateral switches. The same rule about an active 0 state applies (common terminal grounded). In the circuit of Fig 3.4.2 the thumbwheel setting is subsequently scanned by conventional multiplexers. This explains the use of the negated form of thumbwheel coding (denoted by $\overline{1}\ \overline{2}\ \overline{4}\ \overline{8}$) which turns an active 0 state into an active 1 state.

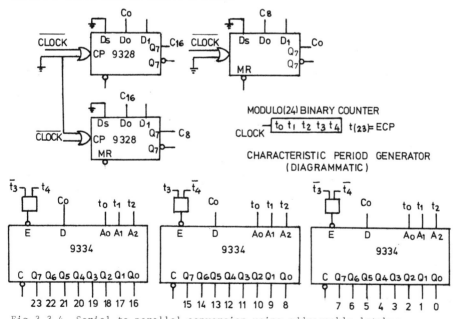

Fig 3.3.4 Serial to parallel conversion using addressable latches.

3.3.5 Latches

Latches are generally made up of D-type Flip Flops (see Fig 3.1.4). A typical device is the dual 4-bit latch type 9308. The data entries at D determine the corresponding outputs Q, while the ENABLE (E) is high. When the latter goes low the outputs remain in their previous state and are un-affected by subsequent changes at the data inputs.

A typical application for the device 9308 is in an 8-channel tape reader. The data inputs are tied to the outputs of photocells illuminated through the perforations in the tape. Each column of holes is accompanied by a sprocket hole of smaller diameter. Its output indicates that the holes corresponding to the current character are safely situated above the photo-cells. If the output from the sprocket hole gives ENABLE, the device will read and store the successive characters on the tape.

3.4 EXAMPLES

3.4.1 Multiplication

The multiplication superprogram for $d \leftarrow b \times c$ is given in circuit form in Fig 3.4.1. b and c are loaded into the 16 bit registers before the sub-program is entered $(\bar{\alpha})$. The product d is formed in the 32 bit register. The program follows the algorithm given in Table 2.3.1.

3.4.2 BCD-binary conversion

The complete circuit for BCD-binary integer conversion is given in Fig 3.4.2. The input data is held in 6 BCD coded thumbwheels. The common terminal of the thumbwheels is earthed. The negated coding $(\bar{1}, \bar{2}, \bar{4}, \bar{8})$ gives the data in the form of Fig 2.3.1. The holding register (C) is assumed to have a 22 bit resolution. The program follows the algorithms given in Table 2.3.3.

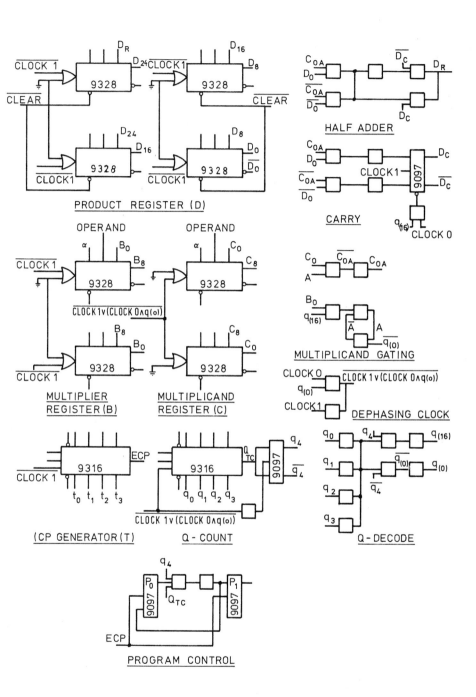

Fig 3.4.1 Complete circuit for multiplication subprogram (16 bit resolution).

104

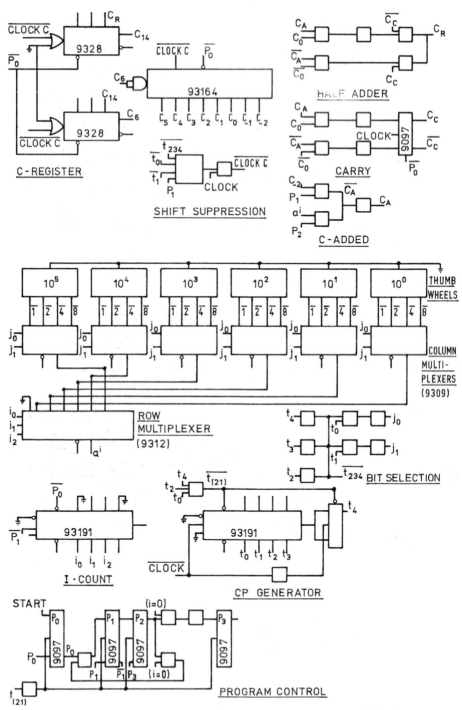

Fig 3.4.2   BCD-binary conversion of 6 decade input from thumbwheels.

4   PRACTICAL CONSIDERATIONS

4.1   PROPERTIES OF DIGITAL INTEGRATED CIRCUITS

4.1.1   General considerations

As a glance through manufacturers' literature will show, digital integrated circuits .(ICs) may be grouped into 'families' of which there is a large and increasing variety. Each family has particular properties and the variety proves that there exists no universal ideal.

Most of the logic design described here will be discussed in terms of a group of broadly compatible devices which is called CCSL (Compatible Current Sinking Logic). This comprises both the DTL (Diode Transistor Logic) and TTL (Transistor Transistor Logic) logic families, whose properties are considered in Section 4.1.2. (For greater detail on the available devices reference should be made to the manufacturers' data books. A compendium [4.1] is also available.)

CCSL devices are at present still the most widely used type of IC. All the basic forms of gating and Flip Flops are available in both DTL and TTL. The complexity of TTL devices goes much further, so that compound logic functions are available in a single package. This feature is called MSI (Medium Scale Integration) and the number of gates per 'chip' can be between 10 and 100. More complex devices still, called LSI (Large Scale Integration), are provided for more specialised logic functions, large scale memories and microprocessors. Mostly they rely on Metal Oxide Semiconductor (MOS) technology.

With some minor restrictions DTL and TTL devices may be readily inter-mixed. Other types of IC - the low-power and high-speed versions of TTL and CMOS (Complementary MOS) devices (described in Section 4.1.7) - are compatible with DTL/TTL but not as readily interchangeable. One useful feature of all these 'families' is the fact that many functions share the same package type and pin connections. Redesign to make use of ICs from a different family is thus minimized.

Three drawbacks of CCSL should, however, be mentioned.

TTL was designed primarily for use in the high-speed logic of computers. In the generally slower processes found in industrial applications, this speed is unnecessary and may be an embarrassment*. High-speed operation inevitably involves switching transients. These can generate considerable electrical interference in adjacent circuits. In turn, the devices are capable of responding to high-speed signals and are therefore vulnerable to unwanted pulse spikes from such interference (see 4.1.6).

The second weakness of CCSL is its relatively large power consumption. While by the standards of the controlled processes power dissipations are small, there are good reasons for keeping power consumption within the control unit low (by using, for example, a low-power version of TTL known as LPTTL or by using CMOS devices). Low-power systems are electrically quieter. Switching transients carry less current, thus reducing problems of mutual interference. In large CCSL systems the total power consumption itself can become significant. Stabilised power supplies capable of delivering several amps of current will then be needed. Lastly, overall heat dissipation can become significant and reliability reduced due to overheating of the IC chips.

The third weakness of CCSL is the low voltage swing (nominally 0 to 5 volts) between the logic 0 and logic 1 states. As a result, should variations in earth potential or other perturbing voltages at the input to the logic element exceed a volt or so, a false logic level may be generated (see 4.1.3 and 4.1.6).

Where high speeds are not required it is therefore preferable to use devices which operate over a much larger voltage swing (0 to 12 or 15 volts). Such devices have been developed from DTL and TTL and are known as High Level Transistor Logic (HTL) or High Noise Immunity Logic (HiNIL). Some versions also have an extra external connection to which 'slow down' capacitors may be attached. This may be particularly useful where input signals from the plant are noisy. HTL-type circuits can also be used to drive much heavier loads than their CCSL equivalents.

An attractive technology for industrial control is a combination of CMOS (see 4.1.7) and HTL which have similar logic voltage swings. They may both be run from the same 12 or 15 volt supply [4.2] . Since CMOS lends

---

*However, even if the controlled process is relatively slow, high speeds may be necessary in certain parts of the controller. For example the internal computations which have to be executed in preparation for controlled action (the computing unit described in 2.7.1 needs a 1 MHz clock in order to complete its computation in 7 ms).

itself to greater complexity of integration and is able to provide relay
type switching functions, one would use it in the more complex control and
calculation sections of the controller. The more robust HTL circuits would
provide the interface with the external world.

In spite of the above reservations, CCSL circuitry will give satisfactory performance,provided attention is paid to the layout of logic
circuitry and other techniques of eliminating noise problems given in this
chapter. In fact the problems associated with CCSL occur to a greater or
lesser extent in other logic families. For this reason the description which
follows will largely deal with CCSL.

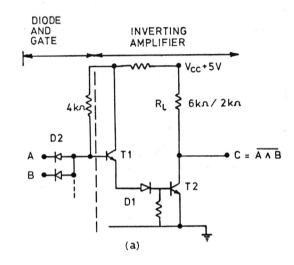

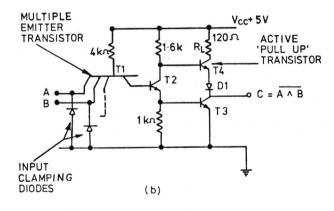

Fig 4.1.1 Basic gates (a) DTL   (b) TTL.

4.1.2   The CCSL family of logic devices

The DTL gate (Fig 4.1.1a) was designed as an integrated version of discrete circuits which used a diode gate followed by an inverting amplifier to give increased output drive and a sharper 'knee' in the transfer characteristics (Fig 4.1.2).

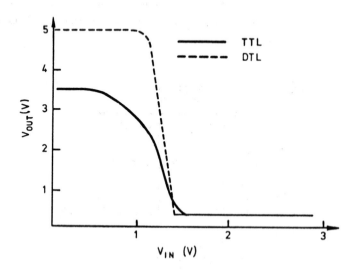

Fig 4.1.2   Transfer characteristics of DTL/TTL gate.

If both inputs A and B are above a certain voltage, diodes D2 are reverse biased and the current through the 4kΩ resistor will flow into transistor T1, switching both transistors on and giving a low output.   If the voltage at A or B drops below the switching threshold, current is diverted from the base of T1 into the input.   Both T1 and T2 then switch off and the output rises to the high state.

The potential V_{TH} at which switching occurs is given by

$$V_{TH} = (V_{bel} + V_{D1} + V_{be2}) - V_{D2}$$

$$= 2V_D \cong 1.3 \text{ volt,}$$

where V_{bel} and V_{be2} are the base-emitter voltages of T1 and T2 respectively, which are assumed to be the same as the forward diode voltage drop V_D.

Since only one or other of A or B needs to fall below V_{TH} to give a high output, the function of the circuit of Fig 4.1.1 is that of the NAND gate defined in Section 3.1.1.

The circuit is called 'current sinking' since approximately 1.6 mA must be sunk from an input if its potential is to fall sufficiently below V_{TH} to be recognized as a low input.

In contrast, the current which must be supplied in the high state is only about 5 μA (the leakage current of the reverse-biased diode D2). If an input is left unconnected, it is equivalent to a high input, although its potential (when observed with a high-impedance instrument) will be $\cong$ 1.8V, the base potential of Tl. However, it is often preferable (particularly in the case of TTL) to tie an unused input to a source of logic 1. This may be the output of an inverter whose input is grounded or the supply voltage (+5V) through a 1kΩ resistor.

Similar asymmetry also appears in the output characteristics (Fig 4.1.3) which show that substantial loads may only be driven in the low state. In this state the output transistor T2 (Fig 4.1.1a) is saturated* (overdriven) and is capable of sinking tens of mA. If, however, too great a load is attached and too much current demanded, the collector voltage (V_{CE}) will rise quickly with increasing collector current (I_C) and the transistor will come out of saturation. The power dissipation ($W = V_{CE}I_C$) will then increase and endanger the device.

To improve switching speed, the TTL gate of Fig 4.1.1b was developed. Instead of a number of input diodes, it uses a single multi-emitter transistor Tl which makes the manufacture of the chip easier. Since an NPN transistor may be regarded as a pair of diodes back-to-back, Tl behaves in the same way as a diode input gate with Dl included. The two equivalent circuits are shown in Fig 4.1.4.

To minimize undershoot and reflection from fast pulse edges, input diodes are diffused into the chip to prevent the input voltage falling below $-V_D$(-0.7V).

An emitter-follower T4 was also introduced to lower the output impedance in the high state. As in the DTL gate, if an input is low, T2 and

---

*Saturation in this context means that the base region in the transistor has an excess of charge carriers. This lowers the voltage between the collector and emitter of the transistor which is required to cause a given output current to flow. The gate output displays a very low dynamic impedance. Operation in the 'saturated region' of the transistor characteristics helps to reduce the output voltage of the gate in the low output state (although it tends to reduce switching speeds, see 4.1.7).

hence T3 will be switched off. T4 is now switched on and the output effectively connected to V_{CC} through the 120Ω resistor. When the inputs are high, T2 and T3 are switched on. The base potential of T4 drops, and this, together with the diode Dl, prevents T4 from switching on.

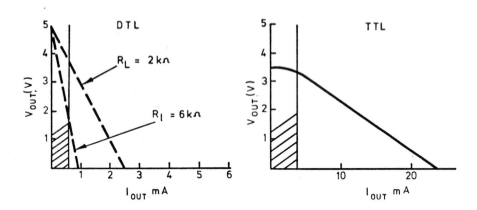

HIGH STATE OUTPUT

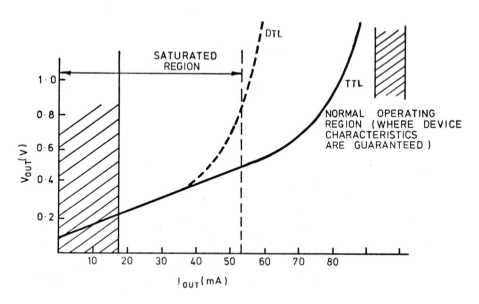

LOW STATE OUTPUT

Fig 4.1.3 Typical output characteristics of DTL/TTL gate.

Unlike the DTL gate, whose unloaded high state output is close to the supply voltage V_{CC} (+5V), that of the TTL gate will be reduced by the voltage drop across T4 and D1 to give $V_{CC} - V_{CE} - V_D \cong 3.6$ volts. The output characteristics are shown in Fig 4.1.2.

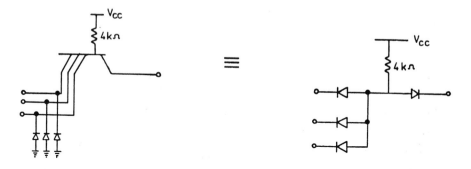

Fig 4.1.4   Equivalence of multi-emitter transistor and diode gate.

Considered as logic functions DTL and TTL gates are equivalent, but with one exception.   The 'Wired-OR' configuration of Fig 3.1.2 is only possible with gates having DTL-type output circuits.   If a TTL output circuit were used, a direct path from V_{CC} to earth would be created between a gate whose output is low and one whose output is high.   TTL gates without the 'active pull-up' transistor T4 and the diode D1 ('open collector' types) are available which allow a passive load resistor to be inserted externally (see Fig 4.1.5).   The determination of load resistor values is explained in the introduction to TTL data books [3.4].

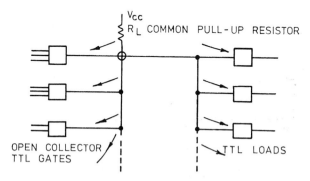

Fig 4.1.5   Wired-OR configuration with open collector TTL gates showing current flow.

Another mechanism, which is introduced in some devices, to allow 'Wire-ORing' is a facility for setting the output of a TTL or CMOS gate into a high impedance state. This is somewhat mistakenly called Tri-State logic. What is really provided, is a separate input which allows one to inhibit the device (make its output impedance high). In CMOS devices it is done by disconnecting the output from both the high and low state voltage supplies; which is easily realised because of the symmetry of the electrical arrangement. Devices with Tri-State output are above all useful where a large number of outputs are to be connected ('Wire-ORed') to a common bus line or highway.

Tri-State outputs also increase maximum switching speeds on a data highway. When a logic '1' is to be transmitted, the active output of the driving gate charges up the capacitance of the bus quickly (since all other connections to the bus have a high impedance).

4.1.3 Loading rules

These rules ensure that logic levels will be maintained within acceptable limits shown in Fig 4.1.3. If too many gates are driven, the output voltage will rise in the low state or fall in the high state. The circuits may then fail to give the correct logic function.

In practice, it is important to guarantee correct functioning in the presence of small changes in voltage levels. The limits of permissible level drifts define the DC noise margin. The bounds of the guaranteed high and low output states for fully loaded devices are illustrated in Fig 4.1.6. The regions between these bounds and the upper and lower bounds of the switching region are called the high and low level noise margins respectively.

Within the DTL family, the standard gate is guaranteed to sink 12 mA without the output rising above 0.4V. Since each driven input requires 1.5 mA, the fan-out in the low state is 8 (Inputs of the same NAND gate driven in parallel provide almost the same load as one input). More than 8 gates can, however, be driven in the high state since the load is only 5 μA/input, which is negligible compared with the leakage current of the output transistor (120 μA).

The TTL gate can sink somewhat more current in the low state (15 mA), giving a fan-out of 10. The increased current available in the high state due to the 'active pull up' in the output stage (∼1 mA) is offset by higher input leakage currents (∼50 μA). The fan-out in the high state is therefore about 15, when the output leakage current is included.

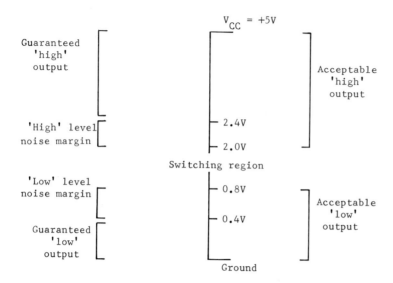

Fig 4.1.6 DTL/TTL standard logic levels.

In 'Wired-OR' configurations there is an additional effect on fan-out capability. If TTL open collector devices with a single pull up resistor are used (Fig 4.1.5) the accumulated leakage currents reduce the fan-out capability in the high state. In DTL, where wired-ORing is permissible with standard gates, R_L will be the harmonic sum of the individual gate load resistors and fan-out capability in both states will be affected. (Tables of permitted fan-in/fan-out combinations will be found in the manufacturers' literature).

Buffer gates are available in both DTL and TTL to drive larger fan-outs of 20-30 gate inputs or other large loads.

Compatibility in the direction TTL $\rightarrow$ DTL presents no problem (fan-out $\sim$10) since TTL can drive more current in either state. A DTL gate can, however, only drive 4-5 TTL gates owing to the combination of high leakage current of the TTL loads and high output impedance of the DTL gate in the high state. The fan-out may be raised to 8 by using gates with a 2kΩ load resistor or by using an additional external resistor of $\sim$2.2kΩ.

4.1.4 Switching speed

Fig 4.1.7 is a diagrammatic representation of the propagation of a pulse through a DTL or TTL gate, where t_r and t_f are respectively the rise time and

fall time of a pulse measured between 10 and 90 per cent of its amplitude.
If the switching point on the edge of the pulse is taken as 1.5V the
propagation delays t_{PD-} and t_{PD+} between the input and output of the gate
are defined as shown.

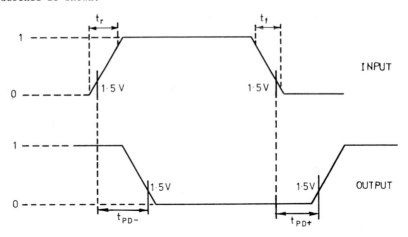

Fig 4.1.7   Pulse propagation through DTL or TTL gate.

The values of the propagation delays depend on external loads but
typical values are 10 ns for t_{PD+} and 7 ns for t_{PD-} in TTL.   DTL is generally
slower giving t_{PD+} = 35 ns (2 kΩ 0/P resistor), t_{PD+} = 45 ns (6 kΩ 0/P
resistor) and t_{PD-} =. 20 ns.

The propagation delays are the result of the internal delays within
the integrated circuit chip and the external delays in the driven circuit.
The former are due to the effects of charge storage in the diodes and transistors and other internal capacitances.   The latter include the input
capacitances of driven gates and the inductance and capacitance of leads and
connecting cables.   The rise and fall times t_r and t_f (Fig 4.1.7) are very
much affected by external capacitances.   In the transition to the low output
state external capacitance is discharged quickly through the low impedance
of the 'saturated' output transistor.   However, in the opposite transition,
this capacitance can only be charged slowly through the load resistor.   This
is particularly so in the case of DTL where the output resistance is high.
Slow rise times (t_r) and rounded rising pulse edges are common in DTL
circuitry.   For a typical wiring and gate capacitance of 100 pF, t_r $\cong$ 500 ns.
However, switching of subsequent gates is much faster than this, as the
switching threshold ($\sim$ 1.5V) is well below the 90 per cent of pulse height

level. Cumulative propagation delays, particularly in the case of DTL, can become significant in long gating chains.

A good indication of the limits on switching speed imposed by t_r and t_f is the maximum toggle frequency of a Flip Flop. These frequencies are given as 50 MHz for TTL and 20 MHz for DTL. It is common practice to classify a logic family by these maximum toggle frequencies (High level Logic HTL, $\sim$3 MHz; Emitter Coupled Logic ECL, $\sim$250 MHz). MOS devices are much slower than bipolar devices. The maximum toggle frequency of CMOS depends on the voltage swing in use, but it seldom exceeds 5 MHz even at maximum supply voltage (15V).

A serious noise problem occurs if the rise time of the input to the gate is substantially greater than the propagation delay through the gate. The phenomenon is illustrated in Fig 4.1.8.

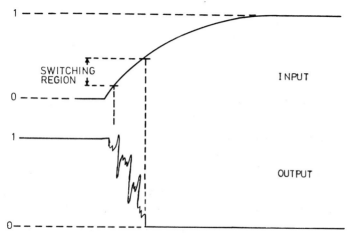

Fig 4.1.8   Instability in the case of $t_r \gg t_{PD}$.

As the input passes through the switching threshold, small amounts of noise can cause oscillations to appear in the output. The time during which the output is in the switching region can be shortened by reducing t_r. This implies reducing the output impedance of the device when the output is in a high state. For DTL, it means simply adding an external resistor in parallel with the device's own output resistor which can provide extra current to charge the load capacitances. It must, however, be remembered that the fanout is thereby reduced.

There are cases where t_r remains excessive or when a long t_r is the inevitable consequence of the circuits being used. A particular case of the latter is the use of low pass filter networks to remove noise from input

signals or to eliminate contact bounce from switch contacts. This problem
may be solved by using a Schmitt trigger circuit to sharpen the edges of the
input signal. The Schmitt trigger is a switching device which has a promi-
nent hysteresis in the transfer characteristic. The triggering level for
switching from the high to low state is different to the level for switching
from low to high (Fig 4.1.9).

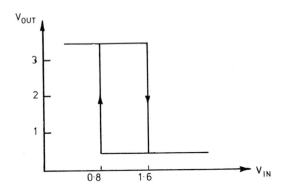

Fig 4.1.9 Hysteresis characteristic of TTL Schmitt trigger gate.

    In the example, if V_{IN} is slowly raised from zero volts, the output
will switch to the low state at 1.6V; V_{IN} must, however, be reduced below
0.8V for the reverse transition to occur.

    Schmitt triggers are available as special TTL gates. Because of the
low turn off threshold, it is important that the circuit driving it must
present a low impedance ($<250\Omega$) in the low state so as to preserve an
adequate noise margin. Special circuits may be constructed which allow a
higher driving impedance or a wider voltage range to be used (Section 4.3.3).

4.1.5 Transmission line effects

    An electrical transmission line consists, in general, of any two
conductors separated by a dielectric insulating medium. It may, for example,
take the form of a pair of parallel wires or a coaxial cable. The current
flow along the line is affected by a distributed series inductance originat-
ing from the magnetic flux surrounding the conductors; the voltage between
them acts across a distributed shunt capacitance. Losses are generally
negligible for reasonably short lines even at the high frequencies used with
DTL/TTL logic.

For a transient waveform travelling along a uniform line, there is a
fixed relationship between the voltage v and current i at all points. This
is called the characteristic transient (or surge) impedance Z_o, where

$$Z_o = \frac{v}{i}.$$

Consider a voltage step or pulse with amplitude V_+ propagating along a
line terminated in resistance R_L (Fig 4.1.10).

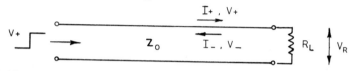

Fig 4.1.10   Transmission line.

When the step reaches the end of the line, v and i must now satisfy the
impedance level set by the termination. If $R_L = Z_o$, there is no conflict;
$V_R = V_+$ and the line will behave exactly as if it extended to infinity.
There is no reflection and the line is said to be matched.

If $R_L \neq Z_o$, the conflicting requirements of v and i can only be
resolved by setting up a reflected signal V_-. Continuity implies that

$$V_+ + V_- = V_R$$

and

$$I_+ - I_- = I_R \quad \text{or} \quad \frac{V_+}{Z_o} - \frac{V_-}{Z_o} = \frac{V_R}{R_L}$$

and therefore

$$\frac{V_-}{V_+} = \frac{R_L - Z_o}{R_L + Z_o}.$$

The reflected signal is therefore inverted when $R_L < Z_o$ and non-
inverted when $R_L > Z_o$. The reflected amplitude and the degree of mismatch
is determined by the difference between the terminating resistance and the
characteristic impedance.

The transmission time for a signal is the time taken for the signal to
travel between two circuits. In practice, a connection between the circuits
need not be considered as a transmission line when the transmission time is
appreciably smaller than the rise time of the signal.

If the transmission time is much less than the rise time, any reflections
caused by discontinuities or poor matching will have time to die away in the
period during which the signal changes state. As the transmission time
increases, multiple reflections start to distort the waveform [4.5]. When
the distortion is severe enough, it can cause false triggering of logic gates.

The internal logic wiring of a control unit may be regarded as made up of transmission lines with Z_o = 100-200Ω. Although the inductance and capacitance will vary somewhat along the length of the wire, it is possible to think of it as a conductor above a ground which is composed of the matrix of the remaining interconnections. Signals travel on such lines with a propagation delay of $\sim$5 ns/m, and so problems should not arise with DTL/TTL gates for line lengths <300 mm.

With longer lengths a single wire may be unsuitable for signal transmission, since the capacitance between the conductor and earth varies abruptly with the position of the conductor relative to other wires and earth. This creates local reflections and will make the characteristic impedance indeterminate. By ensuring more uniform conditions along the line, such as routing the conductor close to a ground plane, the size of these local reflections may be reduced and the maximum safe length increased to $\sim$700 mm.

Extension to longer distances is possible only by employing coaxial or twisted-pair transmission lines which have constant characteristics along their length.

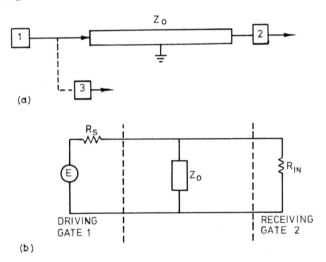

Typical Values

State	E	R_S	Z_o	Connection
High	3.5V	120Ω TTL 6KΩ DTL	50- 90Ω	Coaxial link
Low	0.4V	10Ω	90-120Ω	Twisted-Pair

Fig 4.1.11 Example of logic circuit in which a remote gate (2) is connected by a transmission line of characteristic impedance Z_o
(a) Diagrammatic representation of interconnection
(b) Equivalent circuit for the connection of gates 1 and 2.

Fig 4.1.11a illustrates the case of a logic circuit 1 driving circuits 2 and 3. Circuit 2 is remote and connected by a transmission line of characteristic impedance Z_o. Fig 4.1.11b shows the equivalent circuit for the connection of 1 to 2, where R_s is the output impedance of 1 and R_{IN} the input impedance of 2. Typical values are given in Fig 4.1.11 for the threshold voltages E and the output impedances R_s. R_{IN} is always large.

The basic difficulty with DTL/TTL gates is the matching of the driver and receiver impedances to that of the line. The following points should be noted.

(i) When the output voltage of the driving gate changes, not all of that change is applied immediately to the line. The initial voltage is determined by the potential divider formed by the gate output impedance R_s and the line impedance Z_o. This voltage can only change once the voltage step has arrived back after reflection at the other end of the line. During the high to low transition, $R_s \ll Z_o$, and most of the signal voltage is developed across the line immediately. However, in the low to high transition $R_s \cong Z_o$ for TTL and $R_s \gg Z_o$ for DTL. Therefore the initial voltage at the driving gate may lie in the switching region between 0.8 and 2.5 volts. For this reason, it is generally inadvisable to drive other gates, such as gate 3 in Fig 4.1.11a, from the sending end of a transmission line.

(ii) The input impedance of the receiving gate R_{IN} is much greater than the characteristic impedance of the line and so an incident level change will be reflected at nearly twice its amplitude. The initial step in a high to low transition will bring down the voltage on the line close to zero; the reflection at the gate 2 input will then give a large undershoot. If the receiving gate has input clamping diodes, the amplitude of the reflection will be reduced to less than 1 volt. Further reflections are thereby greatly reduced. By the same mechanism a reflection tending to overshoot the acceptable high voltage level will arise during the low to high transition, but here the incident step is smaller. Overshoot should not therefore occur when using lines of normal impedance ($50-120\Omega$). Overshoot occurring with lines of higher impedance should be limited to $< 5.5V$ to prevent damage to the input transistor of the receiver IC. The likelihood of damage may be reduced by connecting all the receiver gate inputs together, or by using an inverter as the receiver stage.

(iii) When transmitting a low to high transition in conditions where $R_s \gg Z_o$ (ie when using DTL gates as drivers), the voltage will build up in a series of steps as the line charges up gradually by successive reflections. Many transmission delays may be necessary to ensure a high state input to gate 2

with adequate noise margin. During the transition through the switching
region, voltage reflections can cause spurious oscillations in the gate
output as was described in 4.1.4. Terminated lines are therefore recommended
where this is unacceptable.

For matched operation, the cable must be terminated in its character-
istic impedance Z_o - typically 50-90 Ω for coaxial cable, 90-120 Ω for twisted-
pair cable (Fig 4.1.12).

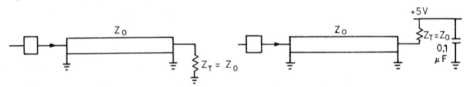

(a) TERMINATION TO GROUND   (b) TERMINATION TO SUPPLY VOLTAGE

Fig 4.1.12 Terminated lines.

A matched termination to ground (Fig 4.1.12a) cannot, however, be
driven by a DTL/TTL gate, as the current drawn from the high state by the
termination Z_T is too great. More current can be sunk in the low state and
a termination to the supply voltage (Fig 4.1.12b) is possible for $Z_o > 100 \,\Omega$
by driving the line with a buffer gate. The two cases are equivalent since
the supply rail is effectively grounded for transient signals by decoupling
capacitors placed across the power supply.

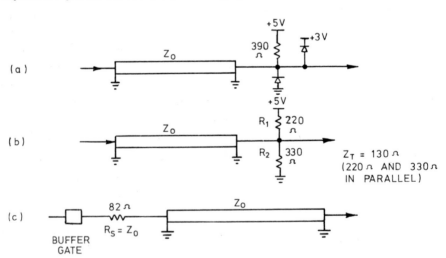

Fig 4.1.13 Termination networks for DTL/TTL signals.

A better solution, as far as power dissipation is concerned, is to terminate the line partially and clip any reflected spikes using diode clamps (Fig 4.1.13a). Since $Z_T > Z_o$, the incident transition will be reflected at a greater amplitude; the diodes eliminate this by preventing the signal exceeding +3.6V or going more negative than -0.6V.

A two resistor network may also be used (Fig 4.1.13b). To a signal on the line, R_1 and R_2 appear in parallel since transient current can run to ground through both of them. Resistor R_1 can then be made larger than the corresponding single terminating resistor, thus reducing power dissipation and the current which must be sunk in the low state.

In some cases, a reverse termination of the reflected signal (Fig 4.1.13c) may be sufficient. The series matching resistor R_s will absorb returning signals and thus prevent multiple reflections at the expense of a small decrease in noise immunity.

During the transient, high frequency currents will circulate at both ends of the line as signals are reflected and absorbed. The supply rail at the driving and receiving ends should therefore be decoupled to prevent noise spikes appearing on the +5V supply to other circuits. A small low inductance capacitor (0.01-0.1 μF) placed as close to the gates or terminations as possible is often sufficient.

By using the above techniques, it is generally possible to transmit DTL/TTL logic levels and pulses reliably over distances of up to 3 metres. Beyond this, noise pick-up may cause problems due to the low noise immunity of DTL/TTL gates. Line drivers/receivers should then be used.

It is not the intention to give the impression that it is always necessary to use terminated lines for the transmission of logic signals. If the receiving circuit is strobed at a time well after the switching transients and reflections on the line have decayed, or if the reflections are eliminated by a low pass filter, unterminated lines need cause no problems. Even if not terminated, coaxial or twisted-pair lines help to provide a measure of protection against externally generated noise.

4.1.6 Noise Immunity

The immunity of a logic circuit to electrical interference can be defined in terms of noise margin - the noise voltage which is required at the input of the circuit to change the state of the output. The d.c. voltage noise margin in the high state is the difference between the minimum acceptable high state input and the guaranteed minimum high state output, the latter

taken at maximum loading. A similar definition applies to the low state noise
margin. For TTL gates (Fig 4.1.6) the guaranteed d.c. noise margin is thus
0.4 volts in both the high and low states.

Generally the d.c. noise margins will be greater than 0.4 volts since,
for example in the high state, the voltages will be much nearer 5 volts than
the guaranteed 2.6 volts. However d.c. noise margins are of limited signifi-
cance.

Noise in logic circuits is really a transient phenomenon resulting from
noise currents flowing in the circuit impedances [4.5]. Typical noise margins
are shown as a function of noise-pulse width in Fig 4.1.14.

Interference may enter the circuit at three points.

(i)   At the input along with the desired signal.

(ii)   On the circuit ground.

(iii) On the power supply line.

Noise is coupled to the input signal by the electromagnetic fields
produced by surges of current in nearby devices. It will appear as an
induced current in the signal conductor. If the wiring to the gate input is
fairly short (not a transmission line), the current may be referred directly
back to the output of the driving gate. The noise source will therefore 'see'
an impedance consisting of the gate input and output impedances in parallel.
Since the input impedance is very large, the induced noise voltage depends
on the gate output impedance. Due to the asymmetry of the DTL/TTL gate the
two states must be considered separately.

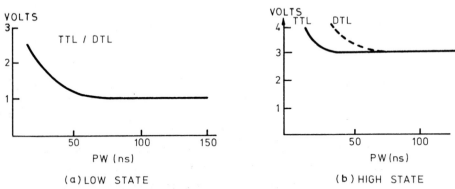

Fig 4.1.14 Typical DTL/TTL noise immunity against pulse width (PW).

In the low state the noise signal must draw more than 50 mA (Fig 4.1.3b)
to produce a perturbating voltage sufficiently large to exceed the typical
d.c. noise margin. As the width of the noise pulse approaches the switching

time of the gate, the pulse will only be propagated through the gate if its
amplitude is greater than the d.c. noise margin. For very short pulses there
is therefore a permissible amplitude (sometimes called the a.c. noise margin)
which is appreciably greater than the d.c. noise margin (Fig 4.1.14a).

In the high state, lightly loaded DTL/TTL gates, have a greater noise
margin than in the low state. However in this state they are more sensitive
to noise because of the much higher output impedance. This is especially so
in the case of the DTL gate where the passive load resistor has a value of
several kilo-ohms.

The conclusion from the above is that noise sensitivity depends on noise-
voltage, circuit impedance and pulse duration. Therefore a more useful
indicator of noise immunity is the minimum energy, $E = VI t$, required to cause
a noise-pulse to change the state of the output. This formulation leads to
Table 4.1.1.

Table 4.1.1 Minimum pulse energy (in nano-Joules) required to change the
output state of a typical gate.

	DTL	TTL
High state	1	2
Low state	4	5

When a noise-pulse current is induced on a transmission line, the
resultant voltage will be determined initially, not by the gate impedances,
but by the characteristic impedance of the line, Z_o. The effect on the
receiving gate will depend on the noise margin given above and the properties
of the transmission line which were discussed in 4.1.5. As a general rule
lines of low impedance will tend to be less sensitive to radiated interference
of this kind than lines of higher impedance.

Ground (or common-line) noise arises from currents which are flowing to
the system. It is easy to give a low resistance path to d.c. currents but,
at very high frequencies, even quite short lengths of wire possess a considerable impedance. Low impedance ground connections can, however, be provided by including one or more ground planes in the circuit construction.
This may be inconvenient where some form of back plane wiring is used. In
that case a rectangular ground mesh can be created by making connections
between adjacent ground pins. A heavy-gauge wire should be used whenever

possible to compensate for the increased impedance of the conductors at high frequencies due to 'skin' effect.

Different sections of the system should be earthed together at a single point which is tied in turn to a 'good' earth in the building. If this is not done, heavy currents drawn by, say power driving circuitry, may influence the potential of the ground connection used by logic circuitry. Fig 4.1.15a shows good and bad arrangements for ground connections. Similar considerations may make it necessary to separate power supply connections to various sections of the controller (Fig 4.1.15b).

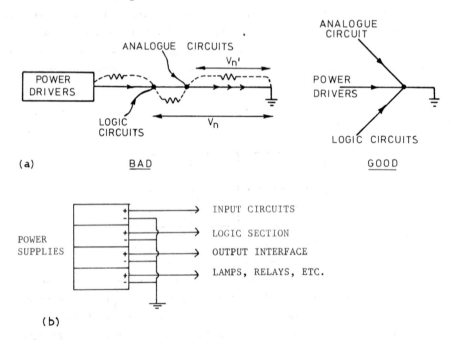

Fig 4.1.15 Application of reference voltages. a) Good and bad grounding arrangements, showing ground noise voltages V_n and V_n' which affect the noise margin of low power circuits. b) Good arrangement of power supplies.

It is characteristic of TTL gates that they tend to generate severe power supply noise. When the gate changes state, both output transistors can be switched on simultaneously for a few nanoseconds owing to slight differences in switching speed. An instantaneous short circuit is thus put across the power supply.

There are two solutions to this problem.

(i) The power supply voltage should be distributed using connecting paths

of low impedance. This can be done by the same techniques used for ground
connections.

(ii) An alternative path to ground for high frequency currents should be
provided close to the noise source. Small low inductance capacitors of
0.01-0.1 μF may be employed. The leads to such decoupling capacitors should
be kept short.

4.1.7 Other compatible types of ICs

Three other forms of digital ICs are directly compatible with DTL/TTL
logic. Their fan-out capabilities differ but logic levels are compatible.

(i)   Schottky-clamped TTL (TTLS)

The chief limitation on the speed of the basic TTL gate is due to
charge storage in the collector-base junction of transistors which are
switched on and held in the 'saturated' state. This charge must first be
removed before the transistors can switch off. If Schottky-barrier diodes
are diffused across these junctions, the transistors are prevented from
'saturating' and the switching delays are considerably reduced
(typically $t_{PD} \cong 3$ ns).

The consequent very high rate of change of voltage (up to 10^9 V/s) can
cause interference problems and the recommendations made in Sections 4.1.5
and 4.1.6 regarding interconnections and decoupling capacitors must be
rigorously adhered to.

(ii) Low Power TTL (LPTTL)

One disadvantage of DTL/TTL which has been mentioned, is its high
current requirement.

By changing component values and circuit construction, low power TTL
reduces the power supply current drawn to 1/10 of that required by the
corresponding TTL circuit. Since the currents switched are correspondingly
reduced, switching time is increased ($t_{PD} \cong 35$ ns). If this is unsatisfactory,
the switching speeds of TTL may be restored by using a Schottky technique.

Power supply noise is much less than with TTL. Problems of interference,
such as crosstalk, are reduced since the power of high frequency disturbances
generated by the switching is much smaller than with TTL.

(iii) Complementary metal-oxide-semiconductor (CMOS)

The operation of the circuit may be explained by means of the inverter shown in Fig 4.1.16. When the input is low, the N-channel MOS transistor will be off and the P-channel on. The output is thus effectively connected to the supply voltage V_{DD} through the P-channel resistance (typically a few hundred ohms). When the input is high, the roles are reversed and the output is grounded through the N-channel resistance. The output impedance of CMOS

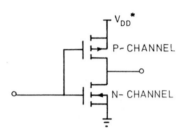

Fig 4.1.16 CMOS inverter (*For compatibility with DTL/TTL, V_{DD} = +5V)

devices is therefore symmetrical. The MOS transistors used, have an extremely high input impedance ($\sim 10^{11}\Omega$) and so very little current is drawn in either the high or low states. Significant dissipation does however occur in the output stage during switching.

CMOS differs significantly from DTL/TTL in the following respects:

1   Very low power requirements

In a quiescent state a CMOS gate dissipates less than 1 μW of power, but during switching dissipation is much greater. The power requirement of a CMOS system is thus dependent on clock frequency. Typical values are given in Table 4.1.2. The power varies linearly with frequency and at a rate greater than the square with supply voltage. It also increases with the capacity of the load.

Table 4.1.2 Power dissipation in CMOS gate (15pF load)

	d.c.	200 kHz	1 MHz
V_{DD} 5 V	25 nW	0.12 mW	0.60 mW
V_{DD} 10 V	50 nW	0.6 mW	3 mW
V_{DD} 15 V	75 nW	1.8 mW	9 mW

2   Wide range of supply voltage (3-18 volts)

Tolerance of supply voltage variation, together with the low power requirements, allows the use of smaller, less well regulated power supplies. In remote areas, battery operation may be possible. The supply voltage can be chosen for compatibility with CCSL (V_{DD} = +5V) or HTL (V_{DD} = +12-15V).

3   CMOS is slower

Signal rise and fall times and propagation delays are much longer than with TTL or DTL. The maximum Flip Flop switching speed (3 MHz at V_{DD} = 5V, 9 MHz at V_{DD} = 10V) is accordingly reduced. Another feature of the waveforms is their nearly equal rise and fall times due to the symmetrical output impedance (200-500ns). Transmission line effects are, as a result, less pronounced and restrictions on layout and interconnection lengths can be relaxed somewhat compared to circuitry using DTL/TTL.

4   Noise immunity

Since the input threshold voltage of CMOS tracks the supply voltage V_{DD}, the d.c. noise immunity is typically around 0.45 V_{DD}. At V_{DD} = 15V, this is obviously many times greater than for TTL or DTL. The a.c. noise immunity will also be significantly better since CMOS gates are slower than their DTL/TTL counterparts.

5   Higher fan-out

The load due to CMOS input is almost neglible except for its capacitative effect (see below). A CMOS gate may drive more than 50 other gate inputs.

The above factors represent convincing reasons for using CMOS, particularly in industrial control applications. There are however a few other problems which should be discussed further.

1   CMOS has a comparatively high output impedance and is much more sensitive to the capacitance of the load than TTL. The limit to fan-out from CMOS devices is therefore set rather by the capacitance of the driven inputs and wiring, than by the current requirements of the gate inputs.

2   As was mentioned in 4.1.6, noise immunity depends not only on noise voltage and noise pulse width, but also on circuit impedance. Here CMOS is at a disadvantage or at least no better than DTL or TTL, since its output impedance is greater.

3   CMOS inputs have an extremely high impedance and it is strongly recommended that their direct connection to wiring passing outside the control logic be avoided. Bipolar devices should be used as buffers.

4   CMOS inputs are generally described as having input protection in the form of diodes diffused into the IC chip. It should be noted that this

protection is only valid for short periods of overvoltage, not exceeding given values. Where external signals may exceed these limits, protective circuits as described later in this chapter should be used. Static charges may also build up when devices are handled and the manufacturers' instructions on precautions should be followed.

In a CMOS circuit all inputs should always be tied to one or other of the supply voltages, so that no MOS gate connection is ever left floating.

The interfacing of IC types is dealt with fully in the manufacturers' literature. Where there are voltage level conversions, interface circuits are available as part of the family.

The other consideration is loading. The answer to the question as to how many devices of one type can be driven by the output of a device from another family is summarised in Table 4.1.3.

Table 4.1.3 Fan-out compatibility of DTL/TTL/LPTTL/CMOS.

Driver	Receiver			
	DTL/TTL	TTLS	LPTTL	CMOS (V_{DD} = +5 V)
DTL/TTL	8/10	8	40	>50[+]
TTLS	12	10	50	>50[+]
LPTTL	1	1	5	>50[+]
CMOS	≠	≠	1	>50

Notes:

d.c. fan-out only is shown. The table applies to standard gates. Buffer gates have increased driving capability.

+ Not recommended to be driven by TTL gate with active output since the CMOS input threshold voltage is above 2.6V.

≠ Special devices are needed to sink the necessary current.

The compatibility of the various forms of TTL devices is evident. The main exception is the need for buffers between the output of a low power device and the inputs of standard devices. A further feature is the pin for pin compatibility of the devices. Thus increasing the switching speed or reducing power requirements can be achieved without rewiring the system. A mixture of various types in a complex control system is often the final solution.

4.2   SIGNAL TRANSMISSION

4.2.1   Introduction

The use of high speed, low level logic signals within the control unit
has the unfortunate consequence that they cannot be used externally to com-
municate directly with the plant. The ambient noise level in many factories
may be an order of magnitude greater than the 0.4V guaranteed noise-margin
for DTL or TTL gates.

Since the outgoing signals generally have to be amplified in order to
drive plant equipment, it could be considered feasible to perform the conversion in the control unit and transmit the high power signals. Even this,
however, is not satisfactory since power-switching transients create unacceptable levels of noise. In general therefore power amplifiers should be sited
away from the control unit and data transmitted to them as low-power signals*.
Similarly,the direct connection of sensors in the plant to the input of
DTL/TTL gates in the control unit is unsatisfactory. Such a link acts as an
aerial for the pick up of noise.

Finally the transmission of analogue signals over long distances should
be avoided where accurate measurements are to be made. Digital signals will
tolerate noise-amplitudes up to the noise margin of the transmission circuits.
The limit for noise on analogue signals is set by the required percentage
accuracy. It is thus more satisfactory to convert the signal to digital form
at the source.

This means that there is a need for special transmission techniques
for low power digital signals which eliminate or minimise interference due
to noise.

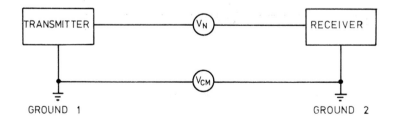

Fig 4.2.1 Noise in single-ended transmission system.

*In extreme cases of noisy environment,transmission using a different medium
such as light (fibre optics) may be necessary [4.6].

It is common practice in the transmission of electrical signals to use a two conductor system consisting of a signal wire and a ground return connection. Noise problems can therefore be divided into two classes (Fig 4.2.1).

(i)   Induced noise on the signal conductor, V_N
(ii)   Difference in potential between the ground references used by transmitter and receiver (common-mode noise), V_{CM}.

Induced noise can be reduced by using screened cable. The use of a low impedance matched transmission line will also minimize the noise voltage 'seen' by the receiver. In many cases, increasing the signal voltage swing will be sufficient and filtering may be used to eliminate any remaining high frequency noise spikes. For logic signals a detector incorporating Schmitt trigger action is generally necessary to convert the resulting slow pulse edges into signals which may be used within the control unit.

Common mode noise is unaffected by filter networks or cable quality, since it is induced in both conductors. It can only be tackled by providing isolation between the circuit grounds. Devices such as relays, transformers and optical isolators prevent currents circulating between transmitter and receiver through ground connections and thus avoid ground noise voltages (Fig 4.2.2).

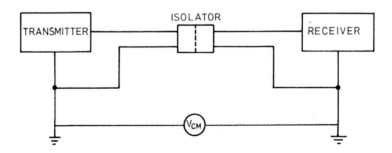

Fig 4.2.2 Removal of ground noise using isolator.

An alternative approach is to provide the transmitter with signal and return outputs which are isolated from the local ground (Fig 4.2.3). If a comparator circuit is used as the receiver it will respond only to the

voltage difference between the two outputs and any noise appearing on both
signals will be rejected.

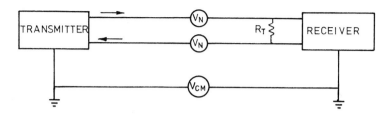

Fig 4.2.3 Removal of ground noise using differential transmission.

Differential transmission is particularly effective when the two signal
wires are close together - as they must be if a transmission line is used.
Any induced noise will then appear almost equally on both wires and thus be
rejected.

Reliable data transmission in high noise areas can be achieved by
using all or some of the following techniques according to the severity of
the problem.

1   Mechanical isolation (4.2.2a)

2   Optical isolation (4.2.2b)

3   Conversion to high level, bipolar, logic signals (4.2.2c)

4   Differential transmission (4.2.2d)

5   Filtering (4.2.3)

6   Attention to choice of cable and layout of connectors (4.2.4)

4.2.2 Digital transmission techniques

(a) Relays

Miniature relays offer a convenient method of controlling external
equipment at low repetition rates and where very fast response is not
required. Reed relays are the simplest and most universal form of buffer to
use. They are available packaged in dual-in-line form and can be driven
directly by ICs. Switching action takes a few millisecs but suffers from
'contact bounce', owing to the mechanical oscillation of the reed contacts,
during this period. Mercury-wetted types avoid this problem but must be used
in a particular attitude. Since the contact area is small, reed relays cannot
switch large powers (10 W is typical) and must therefore be used with

contactors or the semiconductor devices described in 4.4.2 to switch heavy
electrical loads.

The mechanical delay in operation can be an advantage when noisy input
signals are encountered,since the coil must be energized for quite a long
period to cause false operation.

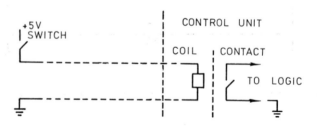

Fig 4.2.4   Relay used to repeat a switch signal.

There is also a very high degree of isolation between input and output
as there is no connection between the coil circuit and contact circuit
(Fig 4.2.4).

Some recommendations may be useful to avoid the most likely problems
to be encountered in using reed relays.   When high voltages and long leads
are driven, the instantaneous power dissipated at the contacts may become
large due to the discharge of the stray capacitance of the lines.   A small
series resistor placed at the contacts will reduce this and thereby increase
contact life.   The high inrush currents which occur in switching on incandescent lamps may also cause damage.

One way of extending relay life in this case, is to allow a stand-by
current (~40 per cent of on-current) to flow through the tungsten filament
to preheat it, so that its resistance rises to a level much closer to the
normal 'on' resistance (Fig 4.2.5).   Resistance R should be chosen to give a
current just below that sufficient to allow the lamp to glow.

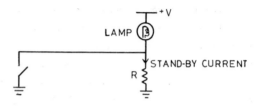

Fig 4.2.5   Incandescent lamps switched by reed relay.

When driving reed relays from ICs, care should be taken to select relay types which include a protecting diode across the relay coil. This will prevent a damaging inductive 'spike' appearing across the coil when the current flowing through it, is interrupted. Otherwise a 0 to 1 transition at the output of the IC would result in a large voltage overshoot on the output transistor. Magnetic shielding may also be advisable so as to cut down radiated noise.

Under normal operating conditions, reed relays can operate successfully for more than 10^8 operations. Since the contact resistance increases towards the end of life, the exact time of failure will depend on the value of contact resistance which can be tolerated in the circuit. In many applications the relay may last considerably longer than the above figure suggests.

Reed relays are probably the most useful interface element. They can be used for both signal transmission and reception and also fulfil an isolating and interference-filtering role. One application would be as buffer elements in the transmission of interlocks between control units (Fig 4.2.6). The delay in signal transmission caused by the relays can be taken care of by the techniques of synchronization explained in Section 2.5.2.

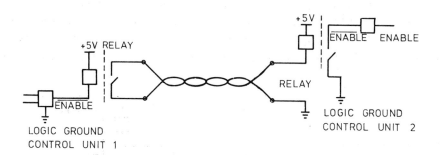

Fig 4.2.6 Transmission of interlock between control units through 2 reed relays.

(b) Optical isolators

Optical isolators consist of a light emitting diode (LED) optically coupled to a phototransistor detector (Fig 4.2.7). With no electrical contact between the two stages, common-mode rejection between the input and output is very high, even at high frequencies. Optical isolators are available in IC form, the detector output being either from an open-collector transistor or a TTL gate.

Depending on packaging, these devices can withstand d.c. voltages up to 2500V between the input and output sections. Propagation delay times range from 50 ns to several μsecs.

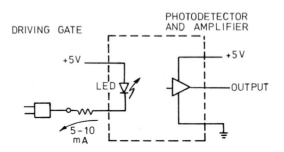

Fig 4.2.7 Typical optical isolation.

In providing local isolation the optical couplers are an alternative to reed relays for low frequency transmission and to pulse transformers for high frequency transmission (up to 5 MHz). Their sensitivity at high frequencies, however, makes their use as a line receiver difficult, since reflections and spikes may cause multiple switching transients.

An application of optical isolators is the elimination of ground loops within the control unit. If interference from relays or power drivers is a problem, optical isolators may be inserted in the signal lines (Fig 4.2.8).

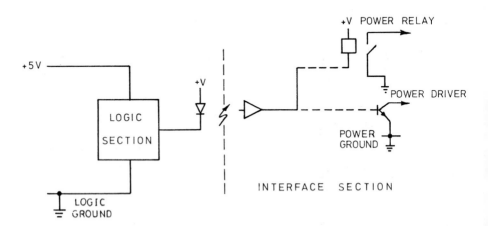

Fig 4.2.8 Optical isolator as buffer between logic and interface.

(c) High level, bipolar CCITT/EIA compatible signals

Integrated circuits capable of transmitting and receiving signal levels
compatible with CCITT/EIA specifications are available from several manufacturers. The logic levels used are nominally ±6V (min.) feeding an impedance
of 3-7 kΩ with

$$\text{logic '0'} < -3V$$

$$\text{logic '1'} > +3V$$

The use of bipolar signals halves the signal power required for a given
output voltage swing. The voltage noise immunity is therefore high (a minimum of 3V).

Other useful features in such line drivers are

(i) protection of output against accidental short to ground or voltage
between say -15 and +15V,

(ii) predetermined output state when the power is switched off,

(iii) slew rate limiting by external capacitor (without a capacitor the output voltage can change at a rate greater than 30V/μs. Such a large voltage
gradient could cause problems of crosstalk between adjacent signal cables).

Useful features in the receiver are

(i) predetermined output when the input is disconnected (open-circuit),

(ii) optional external capacitor for use in signal filtering,

(iii) input hysteresis of up to 1V to cope with rounded pulse edges caused by
filtering.

Bipolar CCITT/EIA signals are excellent for use at frequencies up to
several kHz in areas with fairly low common-mode noise. The slew rate of
the transmitter and filter time-constant of the receiver can be adjusted to
give the best noise immunity in given conditions of data rate and cable
length. The robustness of the devices and components used for transmission
of CCITT/EIA standard signals makes them ideal for operation in industrial
environments.

(d) Differential transmission

As mentioned above, another way of reducing the effect of both induced
and ground noise is to isolate signal and return conductors from ground and
detect the difference in voltage between them.

Several ICs are available for this purpose. The line drivers are
generally constant current devices operating in one of two modes.

(i) Push-pull, where one input sinks and the other supplies current in one
state and their roles are reversed for the other state.

136

(ii) Current switching where a constant current is switched into one or other of the line conductors according to the logic state.

(i) has the advantage of providing double the receiver voltage swing for a given current, (ii) has lower power requirements and may be used in party line systems since a third, inhibit state (where no current flows in either line) is available.

A corresponding range of receivers is manufactured. They are differential comparators with a very narrow dead-band and provide a TTL/DTL compatible output logic level (Section 4.3.3).

A typical transmission system is shown in Fig 4.2.9.

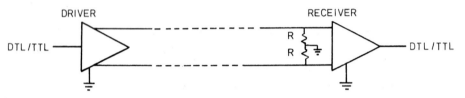

Fig 4.2.9 Differential transmission.

Some form of load is required for the driver outputs. This is supplied by matched symmetrical resistors R, as in Fig 4.2.9, which present an impedance of 2R to differential signals and R/2 to common-mode signals. The line will therefore be matched if $R = Z_o/2$. The common-mode impedance will then be $Z_o/4$ which increases the common-mode noise immunity.

If the voltage applied simultaneously to both input terminals of the receiver exceeds a certain limit (the maximum common-mode voltage), normal operation cannot be guaranteed. This range is approximately ±3V for most circuits. However, it can be extended by the use of input attenuating networks at the expense of reducing sensitivity. If a 5 : 1 attenuation is used and receiver gain is high enough, ±15V common-mode signals can be rejected (Fig 4.2.10).

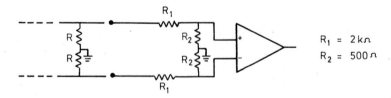

$R_1 = 2k\Omega$
$R_2 = 500\Omega$

Fig 4.2.10 Extending common-mode rejection to ±15V.

When the driver is disconnected or without power, the receiver can be made to take up a known output state by using bias resistors as in Fig 4.2.11.

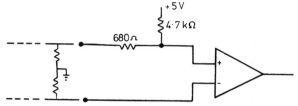

Fig 4.2.11 Biasing receiver to give known output state when disconnected.

The currents switched by the drivers do not as rule exceed 10-20 mA. Differential voltages on the line are therefore about 1V. The low voltages help to reduce crosstalk between lines. If in addition twisted pairs are used, problems from this source can be eliminated in most cases.

The method gives reliable transmission at high data rates over long distances in noisy environments.

4.2.3 <u>Filters and passive elements</u>

It may be necessary to transmit DTL/TTL logic levels over short distances between the control unit and some external source, such as a digital voltmeter or encoder. In this case induced noise, in particular crosstalk between signal conductors in a multicore cable, can be a problem. The simple filter shown below (Fig 4.2.12) will often be sufficient to eliminate the noise $\left[4.5\right]$.

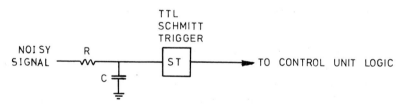

Fig 4.2.12 Hash filter.

The resistor R should be $<250\Omega$ to ensure full noise immunity. Capacitor C must then be chosen to give a time constant (CR) long enough to eliminate interference but short enough to pass information at the required data rate.

Where the noise frequency is close to that of the data a higher order filter with a sharp cut off such as a Chebyshev filter, is needed to reject the noise frequencies $\left[4.5\right]$.

Common-mode noise may also be dealt with by using inductances. This

requires that both signal and return conductors are wound together round a
toroid of high permeability material. The common-mode signal then 'sees' an
inductance proportional to the number of turns and the permeability of the
core. Particularly at high frequencies, this represents a considerable
impedance to circulating common-mode currents. For the differential signals
on the other hand the conductors do not enclose any magnetic material and no
additional inductance is added in their signal path.

Both coaxial and twisted pair cables may be used. A ferrite or
Permalloy ($\mu > 10^5$) toroid ($\sim$100 mm in diameter) is very convenient as a core.
A dozen turns are normally sufficient to provide an inductance of 10-30 mH.
This method may even give some suppression of ground loop interference at
mains frequencies.

4.2.4 Choice of cable and layout of connections

The following types of cable are useful in particular applications.

(a) Coaxial cable (Z_o = 50-90Ω)

This is rather difficult to drive due to its low impedance,but the
outer braided conductor provides good shielding against external fields;
thus avoiding crosstalk. Some penetration will exist with a single-layer,
loosely woven braid but a tightly woven or double woven braid will give
nearly perfect shielding.

Another screen may be added to give a triax cable which can be used
in differential transmission and at the same time provide extra shielding.

(b) Twisted-pair cable (Z_o = 90-150Ω)

The higher characteristic impedance of this cable is easier to drive.
Induced noise appears equally in both conductors due to their close proximity
and twisted configuration. The cable is very suitable for differential
transmission. It is also very much cheaper than coaxial cable, which is a
significant advantage for long runs or multi-way connections.

Additional noise protection can be obtained by screening twisted pairs.
An ideal sheath is an aluminium paper screen within an external copper braid
sheath (Twinax).

(c) Strip cable

Strip cable consists of parallel conductors bound flat together. Severe
crosstalk can be expected between adjacent conductors and it is advisable to

ground alternate conductors to reduce this and provide transmission line
qualities ($Z_o \cong 100\Omega$),(Fig 4.2.13).

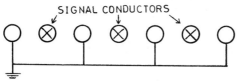

SIGNAL CONDUCTORS

Fig 4.2.13 Strip cable with alternate earth and signal conductors.

Similar flat cable,made from twisted pairs,can be used to give improved
noise and crosstalk characteristics.

(d) Shielded multiway cable

Bundles of conductors within a braided shield have of course, very
poor electrical characteristics and are subject to very severe crosstalk due
to mutual capacitance between conductors. They can, however, be used for
relay signals which are very insensitive to noise and may be used over short
lengths for other signals. Reduction of crosstalk and a more uniform
impedance may be obtained by grounding half the conductors in the cable. A
more effective solution is to arrange the conductors in twisted pairs.

The noise sensitivity of all cable types can be substantially reduced
by employing aluminium or steel cable ducts. Not only does the presence of
a shielding conductor give protection against pickup but the duct itself
may be used to provide a low-impedance ground path and so reduce common-mode
interference.

4.3   SIGNAL PROCESSING

With the introduction of integrated and hybrid analogue circuit devices
much of the detailed design of signal processing circuitry has been eliminated. In the following,the circuit techniques are described in a general
sense. With this basic understanding,a suitable choice of 'black box'
functions can be made by consulting the manufacturers' catalogues [4.4].

4.3.1 Data amplifiers

(a) Introduction

The output of the sensors and transducers used in the control system
is seldom in a form which can be accepted directly by the control logic or

by the analogue to digital converters (ADC) used. One or more of the following
preprocessing stages are needed after the signal source.

(i)   Amplification and scaling of the signal for presentation to level
detectors or ADCs.

(ii)   Buffering and impedance conversion to·avoid the effects of loading the
signal source.

(iii)·Signal Conversion. For example the transducer might rely on a resistance change such as in a temperature measurement. The signal must first be
converted into a proportional voltage or current change for subsequent
processing.

(iv)   Active filters may be needed to separate noise and signal at source.

In all these applications the essential device is the linear operational
amplifier*. The characteristics of a data amplifier should be:

a   High input impedance when used to amplify voltage sources. Low input
impedance for current sources.

b   Low output impedance.

c   Linearity of gain.

d   Constant frequency response over the required range and freedom from
parasitic oscillation.

e   Stability of amplification in the presence of temperature drifts and
ageing.

These characteristics can be achieved, but with difficulty, by special
design of the amplifier for a given application. The general purpose
amplifiers available do not satisfy the requirements when used without feedback. However, external negative feedback improves the characteristics of
the amplifier to such an extent that standard devices can be used for most
applications.

A differential amplifier has two inputs labelled + and - with a large
differential impedance R_D between them. The gain of the amplifier A_o gives
an output V_o which is

$$V_o = A_o(V_+ - V_-)$$

where V_+ and V_- are voltages at the non-inverting and inverting inputs
respectively.

(b)   Non-inverting amplifiers

In the simplest form of feedback amplifier the signal to be amplified

---

*The term operational derives from the use of such amplifiers in analogue
computers to do algebraic operations. In the present context the correct
term would be differential data amplifiers.

(V_1) is at V_+ and a proportion of the output (βV_o) is fed back to the inverting input as V_-. A potentiometer chain of two resistors R_1 and R_2 determines the feedback so that

$$\beta = \frac{R_1}{R_1 + R_2}.$$

The arrangement is shown in Fig 4.3.1a together with the equivalent servo loop representation, Fig 4.3.1b.

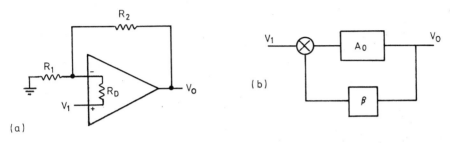

(a)   (b)

Fig 4.3.1 Non-inverting feedback amplifier (a) Circuit representation (b) Equivalent servo loop representation.

From 4.3.1b the relation is

$$V_o = (V_1 - \beta V_o)A_o$$

whence the closed loop amplification is

$$A = \frac{V_o}{V_1} = \frac{A_o}{1 + \beta A_o}.$$

If A_o is large, $A \cong 1/\beta$. This result would follow directly if $V_+ = V_-$, which means that in a high gain amplifier the required V_o corresponds to a negligible differential input. In such an amplifier the voltage at the inverting input always tends to follow V_1.

The current flowing into the internal resistance R_D is

$$I = \frac{V_1 - \beta V_o}{R_D} = \frac{V_1(1 - \beta A)}{R_D}.$$

Hence the closed loop input impedance of the amplifier is

$$R_I = \frac{V_1}{I} = \frac{R_D}{1 - A\beta} = R_D(1 + \beta A_o)$$

which becomes large for a high gain amplifier and reduces the signal load. Improvements in the other characteristics of the amplifier follow from similar considerations. Thus b   the output impedance is lowered $R = R_o \dfrac{1}{1 + \beta A_o}$,

c   any non linearity is reduced $L = L_o \dfrac{1}{1 + \beta A_o}$,

e   the amplification stability is improved $\dfrac{\Delta A}{A} = \dfrac{\Delta A_o}{A_o} \dfrac{1}{1 + \beta A_o}$,

where in each case a high gain has the effect of making the closed loop characteristic an appreciable improvement over the corresponding open loop characteristic.

Characteristic d, the frequency response, needs further consideration. A high gain reduces the dependence of the amplification on the gain, since in the limit it is determined solely by the feedback ratio β. However the gain decreases with increasing frequency. In addition there is an increasing phase shift up to the point where a 180° phase shift has the effect of turning the negative feedback into a positive feedback. This may cause a parasitic oscillation. There is in fact a maximum β or minimum amplification which can be used with an uncompensated operational amplifier.

A typical dynamic response curve for the amplifier with a maximum gain of 100 dB ($A_o = 10^5$) is shown in Fig 4.3.2.

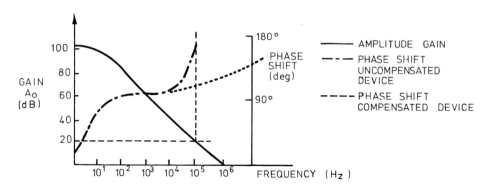

Fig 4.3.2 Open loop characteristic of operational amplifier.

The frequency characteristics of the amplifier are determined by the limitations of transistor frequency response and internal stray capacitances. A point is reached, in an amplifier with the characteristic of Fig 4.3.2, at a frequency in the region of 10^5 Hz, where there is still a significant gain (20 dB or $A_o = 10$) but the phase shift is 180°. Since the amplification of the feedback amplifier is

$$A = \frac{A_o}{1 + A_o \beta}$$

instability results if the magnitude of the loop gain $|A_o\beta| \geqslant 1$, which in the example means $\beta > 1/10$. Hence the minimum amplification (in the low frequency range) which is permissible is $1/\beta = 10$ (20 dB). In other words the closed loop amplification cannot be made smaller than the gain of the amplifier at the $180°$ phase shift frequency.

To obtain stable operation with greater feedback, either the feedback ratio can be made frequency dependent or the open loop gain can be made to fall more rapidly with frequency. To obtain the latter a capacitor should be connected at an internal point of the amplifier to make the gain fall below unity at the $180°$ phase shift frequency. This ensures stability for all feedback ratios.

Contacts for the connection of compensating capacitors are provided on the devices and suitable values of capacitance are suggested in the manufacturers' literature. Also many devices are internally compensated (Fig 4.3.2) so that the problem of instability does not arise.

(c) Inverting amplifiers

An alternative and very common way of using the amplifier is in the inverting configuration of Fig 4.3.3.

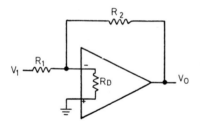

Fig 4.3.3   Inverting amplifier configuration.

As before

$$\beta = \frac{R_1}{R_1 + R_2} \, .$$

The differential input is

$$V_- = V_1 + \beta(V_o - V_1) = \beta\left(\frac{R_2}{R_1} V_1 + V_o\right),$$

the output is obtained from

$$V_o = -A_o\beta\left(\frac{R_2}{R_1} V_1 + V_o\right),$$

and the amplification is

$$A = \frac{V_o}{V_1} = \frac{-A_o \beta \frac{R_2}{R_1}}{1 + A_o \beta} = -\frac{R_2}{R_1} \left(\frac{1}{1 + \frac{1}{A_o \beta}} \right).$$

If the amplifier gain is large $A \cong -\frac{R_2}{R_1}$. This corresponds to the situation in which V_o is the result of negligible current flow into the amplifier and $V_- \cong V_+$. For this reason the inverting input of the amplifier is often called a virtual earth point.

Whereas in the case of the non-inverting amplifier, the input impedance R_I can be very large when the gain is high, there is a limit for R_I in the case of the inverting amplifier.

Using the definition

$$R_U = \frac{R_2 R_D}{R_2 + R_D}$$

for the parallel feedback and differential resistances,

$$R_I = R_1 + \frac{V_-}{\frac{V_-}{R_D} + \frac{V_- - V_o}{R_2}} = R_1 + \frac{1}{\frac{1}{R_U} + \frac{A_o}{R_2}} = R_1 + \frac{R_U}{1 + A_o \frac{R_U}{R_2}}$$

As A_o increases $R_I \cong R_1$. It follows that for high impedance applications large absolute values of R_1 and R_2 are needed. If moreover the required amplification is large, R_2 can easily become excessive. One solution is to use an additional potentiometer chain in the feedback network as in Fig 4.3.4.

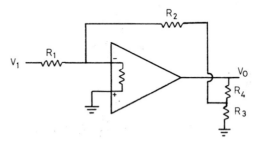

Fig 4.3.4 Inverting amplifier with enhanced input impedance.

Defining the potentiometer ratio $\quad \gamma = \frac{R_3}{R_3 + R_4}$

$$V_o = - A_o \beta \left(\frac{R_2}{R_1} V_1 + \gamma V_o \right)$$

from which

$$A = - \frac{R_2}{R_1} \frac{1}{\gamma} \left(\frac{1}{1 + \frac{1}{A_o \beta \gamma}} \right)$$

so that in the limit (A_o large) $A = - \dfrac{R_2}{R_1} \dfrac{1}{\gamma}$.

The value of R_2 required for a given input impedance and amplification is reduced by a factor of γ .

So far the circuits described have been used to amplify voltage signals and emphasis has been placed on large input impedances. If the signal is a current source, a differential signal could be produced by allowing the current to pass through a load resistor. This, however, becomes impractical if small currents have to be amplified. The load resistor then becomes comparable with the input impedance of the amplifier and the effective load can be influenced by drifts and leakage currents in the amplifier.

The ideal current amplifier is one which has zero voltage drop across the measuring terminals, the configuration of Fig 4.3.5 approaches this.

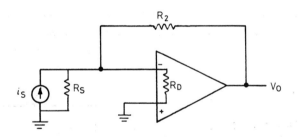

Fig 4.3.5 Current amplifier.

Reverting to the expression for the input impedance (R_I) of an inverting amplifier with $R_1 = 0$, R_D is replaced by the parallel differential and source resistances,
so that

$$R_V = \frac{R_D R_S}{R_D + R_S}$$

and

$$R_W = \frac{R_2 R_V}{R_2 + R_V} \quad \text{replaces } R_U.$$

The potential V_- is now given by

$$V_- = R_I i_s = \frac{R_W i_s}{1 + A_o \dfrac{R_W}{R_2}}$$

and the output by

$$V_o = \frac{-A_o R_W i_s}{1 + A_o \dfrac{R_W}{R_2}}.$$

Therefore for large A_o, $V_o \cong -R_2 i_s$. This corresponds to the situation in which there is no load on the current source due to the amplifier and V_o is developed purely across the feedback resistor R_2.

It must be emphasized that the above analysis is only valid over the region of linear gain of the amplifier. If the differential input voltage rises to the point where V_o is of the order of the supply voltage the output becomes saturated. The amplifier can then sustain appreciable differential voltages and no longer behaves in the manner described.

(d) Bridge measurement amplifiers

Some transducers, such as strain gauges, are mostly used in a bridge configuration. The effects of temperature variations may thereby be eliminated since they affect all four arms of the bridge equally. A differential amplifier (Fig 4.3.6) may be used to measure the out of balance voltage ΔV.

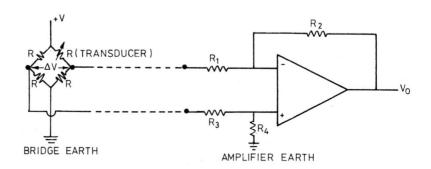

Fig 4.3.6 Bridge measurement using differential amplifier.

If $R_1 = R_3$ and $R_2 = R_4$ it can be shown that

$$V_o = \frac{R_2}{R_1} \Delta V$$

and any common mode voltage, whether d.c. or transient, does not appear at the output. This circuit suffers from two disadvantages:

(i) . for accurate operation the bridge resistances R must be very much less than the input impedance of the amplifier R_I where

$$R_I = R_1 + R_2$$

(ii) resistors R_1 to R_4 must be very carefully matched to obtain good rejection of common mode signals.

The measurement is only meaningful if the amplifier ignores any common mode voltage from the bridge. It should also have good rejection of common mode noise.

For these reasons an instrumentation amplifier is generally to be preferred for such applications.

The output of the above circuit is a non-linear function of the transducer resistance change. When the output is only required to give a null indication this is unimportant. Where a linear response is wanted the circuit of Fig 4.3.7 may be used.

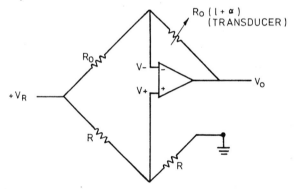

Fig 4.3.7 Linear bridge measurement.

In this configuration the voltages at the input terminals of the amplifier are

$$V_+ = \tfrac{1}{2} V_R$$

and

$$V_- = V_o + \frac{(V_R - V_o) R_o (1 + \alpha)}{R_o + R_o (1 + \alpha)} = \frac{V_o + V_R (1 + \alpha)}{2 + \alpha}$$

since $V_+ \cong V_-$ for a high gain amplifier

$$V_o = -\tfrac{1}{2} \alpha V_R.$$

148

The output is thus linearly dependent on the fractional change in the transducer resistance.

(e) Defects in amplifiers

Up till now only ideal amplifiers have been dealt with. The following
departures from ideal characteristics should be considered in the selection
of a suitable amplifier (Fig 4.3.8).

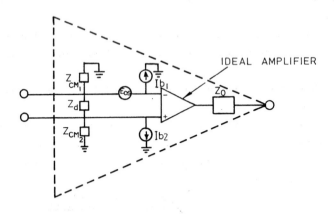

Fig 4.3.8 Real op-amp showing internal impedances, current and voltage
sources.

(i) Offset voltage E_{os}

The input offset voltage (E_{os}) is defined as the voltage which must be
applied between the input terminals to reduce the output voltage to zero. It
exists as a result of imperfect component matching in the circuit and will
vary with temperature*.

In high sensitivity applications (for instance thermocouple measurement),
where E_{os} is of the same order as the input signal, amplifiers with very low
drift ($<1\mu V/^{\circ}C$) should be chosen.

(ii) Input leakage currents I_{b1}, I_{b2}

These arise chiefly from the leakage and bias currents of the input
transistors. They only matter where high input impedances are needed and
hence input and feedback resistors have large values. Leakage currents
flowing through these resistors will cause offset voltage errors at the

_____
*Most op-amps have provision for an external potentiometer to trim the output
voltage to zero.

input. However, provided the differential input leakage current $(I_{b1} - I_{b2})$ is not too large,equalizing source impedances at each terminal will give cancellation.

Fig 4.3.9 shows the introduction of a compensating resistor R_c and the flow of leakage currents. The expressions for R_c,in the case of inverting and non-inverting amplifiers, obtained by assuming $V_1 = V_0 = 0$, are also given.

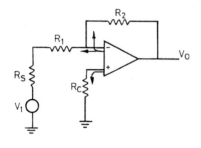

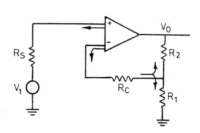

a) Inverting amplifier   b) Non-inverting amplifier

$$R_C = \frac{(R_1 + R_s)R_2}{(R_1 + R_s) + R_2}$$

$$R_C = R_s - \frac{R_1 R_2}{R_1 + R_2}$$

Fig 4.3.9 Compensation for the effect of leakage currents.

In high gain applications temperature variation of differential leakage current may be a problem. In these cases amplifiers with field effect transistor (FET) input stage are preferred. Leakage currents are then limited to a few nA.

(iii) Common mode and differential input impedances

The common mode impedances Z_{CM} (Fig 4.3.8) can be very significant in the high impedance non-inverting amplifier configuration, while the differential impedance Z_D is particularly important in the current amplifier configuration of Fig 4.3.5. These impedances are strongly dependent on temperature and can be the limiting factors on the sensitivity and accuracy of the amplifier.

(iv)   Common mode voltage limit

While a common mode voltage applied to both inverting and non-inverting inputs of the amplifier should have no effect on the output, it could, if large enough, take the input transistors into the non-linear region of their characteristic. The maximum common mode voltage is defined as the voltage applied to both input terminals which will produce a certain percentage error in the input-output relation of the amplifier.

(v)   Common mode rejection

The common mode rejection ratio (CMRR) is a measure of how well a given amplifier will perform in differential measurement. Because of slightly different gains between the inverting and non-inverting sides of the amplifier, common mode voltages will not cancel entirely. The resultant output error ε_o is referred to the input as ε_o/A_o. If a common voltage V_{CM} causes the error, the rejection ratio is defined as

$$\text{CMRR} = \frac{V_{CM}}{\varepsilon_o} A_o .$$

The value is quoted by manufacturers for a d.c. common mode voltage and will be significantly smaller at higher frequencies. It should also be remembered that the criterion applies to the amplifier alone. The quoted CMRR may thus not be attained in practice since mismatch of external components can have an even greater influence.

It is for these reasons that the use of instrumentation amplifiers is often better in high accuracy applications. These contain several amplifiers connected in a closed loop configuration, and one external resistor determines the amplification.

(vi)   Bandwidth and slewing limits

The above defects of amplifier have been described in terms of departures from the ideal d.c. characteristic. It is often important, particularly in the processing of a.c. signals described in the next section, to consider a.c. characteristics.

The bandwidth of an amplifier is usually quoted as the frequency (f_o) at which the open loop gain becomes unity. Operational amplifiers will generally not respond to large signals as fast as such a specification suggests. Another criterion is therefore needed. This is called the slew rate (s) given in volts/s. To relate this to frequency the full linear response frequency (f_p) is sometimes quoted. A sinusoidal oscillation between the output limits (+10 volts) at frequency f_p will have a maximum gradient $(10 \ (2\pi f_p))$ equal to the slew rate.

A typical amplifier might have the following specifications: f_o = 1 MHz, s = 0.5 v/μs. The latter limits the full linear response frequency to 10 KHz.

Apart from this, all the above features of amplifiers must be reconsidered where a.c. signals are concerned. In particular parasitic capacitances, in parallel with input resistances, will reduce their impedance considerably at higher frequencies.

4.3.2 A.C. Signal processing

(a) Amplification

The use of a.c. signals is attractive since narrow band-pass filters can be used to remove noise very effectively and a.c. coupling removes offset and d.c. errors from amplified signals.

The basic inverting and non-inverting d.c. amplifiers of section 4.3.1 may easily be converted to a.c. use by isolating capacitors C_1 and C_2 as shown in Fig 4.3.10.

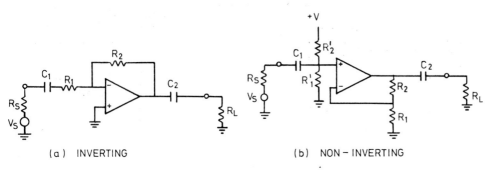

Fig 4.3.10 A.C. Amplifiers.

The combinations of C_1 with the input impedance and C_2 with the load impedance form high-pass filter networks for the incoming and outgoing signals. It is therefore important to ensure that they will allow the signal frequencies to pass without significant attenuation.

C_1 and C_2 must often be quite large and electrolytic capacitors may have to be used*. The latter may only be used in conditions where the a.c. signal is superimposed on a larger d.c. voltage. The d.c. conditions of the amplifier must then be chosen to give a d.c. output of about one half the

---

*Capacitors with paper or plastic dielectric are always preferable where components of suitable values are available. They are unpolarized and have superior high frequency characteristics.

maximum a.c. peak-to-peak voltage. A resistor network,such as $R_2' - R_1'$
in Fig 4.3.10b,would provide this.

(b) Rectification

The control unit cannot accept a.c. signals directly. Therefore one
solution is to rectify and filter the signal to provide a d.c. output for
use with a level detector (Section 4.3.3) or analogue to digital converter
(Section 4.3.5).

If a conventional rectifying circuit were used,an error due to the
forward voltage drop in the diode would be introduced. Therefore a better
method is to use the diodes in the feedback loop,as in the amplifier circuit of Fig 4.3.11.

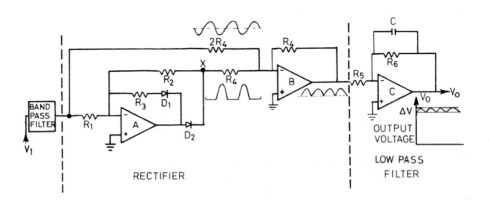

Fig 4.3.11 A.C. signal rectification.

The input signal V_1 is first filtered to remove any noise picked up in
transmission. This is particularly important with low level signals.

The rectifier consists of two amplifiers. Amplifier A acts as a half-
wave rectifier. For a positive input, diode D_1 will be biased on and D_2
biased off. This disconnects the output (point X,Fig 4.3.11) from the amplifier, leaving it connected to the virtual earth point of the amplifier
through resistor R_2. The reverse applies when the input is negative, so
that D_1 is off and D_2 on. The feedback path through R_2 is then re-established
and the output is the amplified input signal. $V_X = -V_1 R_2/R_1$. Since D_2 lies
within the feedback loop, the contribution which the voltage drop across
it (V_D) makes to the output error is effectively V_D/A_o.

Any value of gain may also be obtained in this stage, but unity gain, $R_1 = R_2$, is usually used.

Amplifier B combines the half-wave rectified and original signals in the ratio 2:1 to provide full-wave rectification. For accurate rectification, it is obviously important to have well matched resistors.

The final stage integrates the full-wave rectified signal to give a d.c. output voltage. The amount of residual ripple (ΔV) which appears on the output will depend on the values of R_6 and C. Large values will reduce the amplitude ΔV at the expense of the speed of output response to changes in a.c. signal amplitude.

The shortcoming of the rectification scheme of Fig 4.3.11 is its inability to distinguish between a signal which is in phase and one which is in antiphase with the transducer reference. To process the signal from a bipolar measurement, a phase sensitive rectifier is needed. The transducer reference must obviously take part in the discrimination.

The essential element in the phase sensitive rectification is the analogue multiplier*. This gives an output voltage proportional to the product of its two input voltages. It is used in the configuration of Fig 4.3.12.

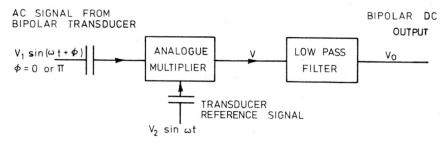

Fig 4.3.12   Phase sensitive rectifier.

The output of the multiplier is given by

$$V = \frac{1}{K} V_1 V_2 \sin \omega t \sin (\omega t + \phi).$$

This has a d.c. component $V_o \propto V_2 \cos \phi$, which is equal to $\pm V_2$ depending on whether $\phi = 0$ or π.

The accuracy of the circuit depends on the accuracy of the multiplier, which is generally quoted as ± 1 per cent. The a.c. component is of frequency 2ω and is removed by the low pass filter. The same comments about the choice of the right filter time constant for a particular application apply as in the circuit of Fig 4.3.11.

*Analogue multipliers are available as hybrid circuit modules $[4.4]$.

(c) Filter amplifiers

Active filters are the result of replacing the feedback resistors of an a.c. amplifier by complex impedances. A simple example of a band-pass filter is shown in Fig 4.3.13.

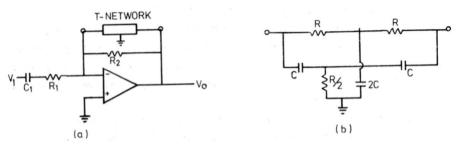

Fig 4.3.13 Active band-pass filter (a) Amplifier configuration (b) 'Twin-T' filter network.

The 'Twin-T' network has a high impedance at a frequency f_o given by

$$f_o = \frac{1}{2\pi RC} \quad ,$$

thus giving a peak amplification at f_o of $- (R_2/R_1)$. The value of coupling capacitor C_1 should be large enough to pass the frequency of interest.

For optimum selectivity and minimum phase shift around f_o both R and R_2 must be chosen carefully. Information on this and the whole subject of active filtering is dealt with exhaustively in the manufacturers' literature.

The advantage of active filters over passive networks is that signal amplitude is maintained and that for low frequency applications, the use of large inductors and capacitors can be avoided.

(d) Phase measurement

Measurement of the phase difference between transducer reference and output signal is a very common way of extracting metering information from a.c. signals. It requires no rectification, is independent of amplitude and can be done digitally with considerable accuracy.

The phase detector consists essentially of a Flip Flop set by the zero cross-over of the leading edge of the reference and reset by the zero cross-over of the trailing edge of the signal*. Agreement in phase then means equal mark space ratio in the state of the Flip Flop. The next step is to make the pulses bipolar and regulate their amplitude. Finally an integration

---

*Schmitt trigger circuits are used to square the a.c. wave—forms before taking them into the phase detector.

over an adequate number of cycles gives a d.c. signal which measures the
departure from equal mark space ratio. This output is bipolar and directly
proportional to the phase angle.

Devices for phase measurement based on the above technique are available in integrated circuit form [3.1] . One very powerful way of using the
phase meter is in conjunction with a voltage controlled oscillator and
variable-modulo counter (see Section 3.3.2) to give a phase locked loop.
The configuration is shown in Fig 4.3.14.

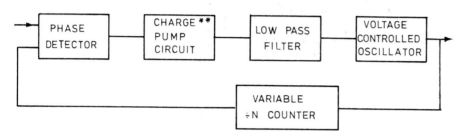

Fig 4.3.14 Phase-locked loop.

The voltage controlled oscillator is set by the phase error to give
a frequency such that N pulses at this frequency are exactly equal to one
period of the incoming oscillation. This means that the period of the input
signal can be subdivided into a fixed number of intervals and this number (N)
can be set digitally.

A phase measurement from an a.c. transducer requires two phase lock
loops (PLL1 and PLL2) whose inputs are the reference and data signals
respectively (Fig 4.3.15).

The outputs of the two phase lock loops are synchronized by a two
phase clock of the same frequency as the reference, so that there can be no
clash of up and down clock pulses. The output of PLL2 requires in addition
a bistable to hold the occurrence of a pulse which, due to phase shift, may
not coincide with clock odd. The least significant digit of the UP/DOWN
counter is ignored since it will oscillate even where there is phase agreement.

A constant rate of phase change alters the frequency of an a.c. signal.
Thus the measurement is really a frequency modulation. The response of PLL2
limits the rate of frequency change which can be detected.

――――――
**The function of the charge pump circuit is to regulate the amplitude of
the pulses.

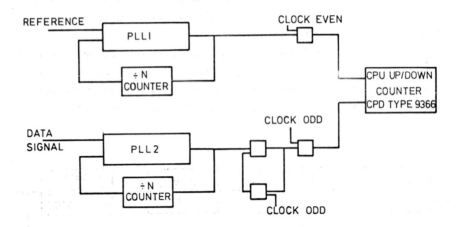

Fig 4.3.15   Digital differential phase meter.

4.3.3   Level Detectors

In many control applications one requires a voltage comparison amplifier
The output of such an amplifier has two states, which indicate whether the
voltage of a data signal is greater or is less than a given preset value.
The characteristics of a comparator are illustrated in Fig 4.3.16.

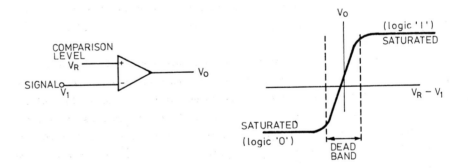

Fig 4.3.16   Voltage comparator.

The comparator is basically a very high-gain differential amplifier
which is operated in the saturated region of its characteristic.   Two factors
are of especial importance.

(i)   Voltage gain: This determines the width of the dead band. As the input voltage V_1 passes through the region where $V_1 \cong V_R$,the output will be in an indefinite state.

(ii) Common-mode and differential mode range: If the differential or common-mode voltage exceeds given limits for the amplifier, the characteristic of Fig 4.3.16 cannot be guaranteed. Damage may even be done to the amplifier. It determines the range of voltages which may be compared and the need for attenuators or protective circuits.

Any suitable amplifier may be used. If special types which give DTL/TTL compatible outputs are not available, the circuit of Fig 4.3.17 can be used to convert the normal operational amplifier output (±10V, ±10 mA) to the required logic levels.

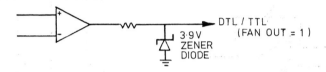

Fig 4.3.17 Conversion of differential amplifier output into logic signal

The Zener diode prevents the voltage rising above 4 volts in the high state and clamps the output to ground in the low state.

The reference voltage must obviously be stable and free of noise to give reliable operation. The source impedance should also be small compared to the impedance of the comparator input. Otherwise current drawn by the comparator will cause V_R to drop and the switching level will vary. Suitable circuits for the reference voltage are given in Fig 4.3.18.

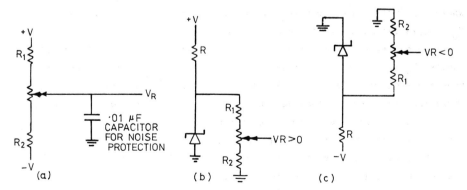

Fig 4.3.18 Voltage reference circuits.(a) — suitable for stabilised power supplies,(b) — stabilised positive V_R,(c) — stabilised negative V_R.

If power supply voltages are stabilized and no great stability of switching level is needed, the circuit of Fig 4.3.18a will be adequate. R_1 and R_2 are chosen to give the required range of V_R. Otherwise a Zener diode may be used to provide a stable potential. A temperature-compensated diode should be chosen and R determined to give a minimum temperature coefficient of zener voltage.

A problem may arise with slowly varying input signals, since the output is undefined when the input voltage is in the 'dead band' region. The solution, as in Section 4.1.4, is to introduce hysteresis into the comparator characteristic. This is obtained by the positive feedback circuit of Fig 4.3.19.

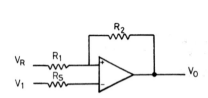 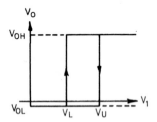

Fig. 4.3.19 Positive feedback comparator circuit for Schmitt trigger action

The upper and lower switching levels are respectively

$$V_U = V_R + \frac{R_1}{R_1 + R_2} (V_{OH} - V_R)$$

and

$$V_L = V_R + \frac{R_1}{R_1 + R_2} (V_{OL} - V_R)$$

where V_{OH} and V_{OL} are the high and low output voltages. The separation $V_U - V_L$ is determined by the choice of R_1 and R_2.

To minimise the effect of offset and leakage currents at the amplifier inputs, R_s is introduced to match the source resistances, as in the case of non-inverting amplifiers (4.3.1). Attention may also have to be given to variations in V_o due to loading since this can affect the switching levels.

Window discriminators* may be made using two comparators to give a 'within limits' indication, as in Fig 4.3.20.

Comparator circuits for plant signals should be near the voltage source. Only digital signals need then be transmitted.

---

*Window discriminators are available as special IC devices.

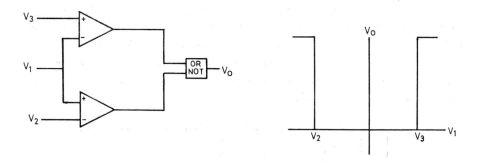

Fig 4.3.20 Window discriminator.

4.3.4 Digital to analogue converters (DAC)

DACs provide an output voltage or current proportional to the value of a digital input vector. The most common technique is illustrated in Fig 4.3.21.

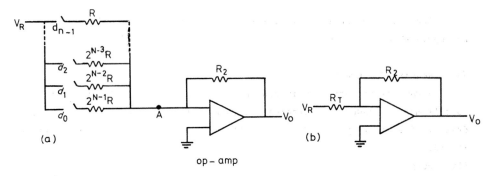

op – amp

Fig 4.3.21 N-bit DAC, showing (b) equivalent resistance R_T.

A switchable resistor network is placed between the reference voltage (V_R) and the summing junction (A) of an operational amplifier. The total current (I) flowing between V_R and the virtual ground of the amplifier is the sum of currents flowing in the parallel resistors. The resistors increase in binary progression from R to $2^{N-1}R$ and are switched into circuit wherever the corresponding bits of the input vector, as shown in Fig 4.3.21, are 1. The output is $V_o = IR_2$ where $I = V_R/R_T$ and R_T is the total resistance corresponding to the input vector $\underset{\sim}{d}$.

The weights of the resistors may be chosen as above, for the conversion of binary input. Alternatively they may be weighted for input in BCD form.

The switches employed may be either relays or solid state. In the latter form, both bipolar and field effect transistors are used. Complete DACs are available as hybrid microcircuits encapsulated in plastic blocks, with resolutions from 6 to 16 bits. It should be pointed out that the resistor network used is in fact of the form shown in Fig 4.3.22, since only resistor values R and 2R are then required.

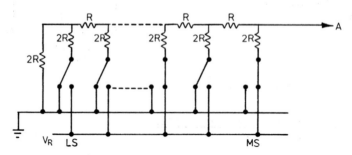

Fig 4.3.22 Resistor network used in hybrid circuit DACs. Node A as in Fig 4.3.21.

The following characteristics should be considered in selecting the appropriate model of DAC.

(i) Required accuracy

A DAC has an intrinsic inaccuracy of $\pm\frac{1}{2}$ the least significant bit (LSB). To this must be added inaccuracies due to resistor mismatch and temperature drifts. The accuracy of a DAC is usually quoted as a percentage of fullscale $\pm\frac{1}{2}$ LSB.

(ii) Output

An operational amplifier buffer may be required to drive the load.

(iii) Settling time

The output of the DAC exhibits damped oscillations when it changes to a new value. The switches may also operate at slightly different times and produce unwanted spikes on the output. The settling time is often given as the time required for output oscillations to decay to a certain percentage of full scale output.

4.3.5 Analogue to digital conversion (ADC)

The level detector (Section 4.3.3) can be viewed as a 1 bit analogue to digital converter. This can be extended to 2 or more bits as in Fig 4.3.23.

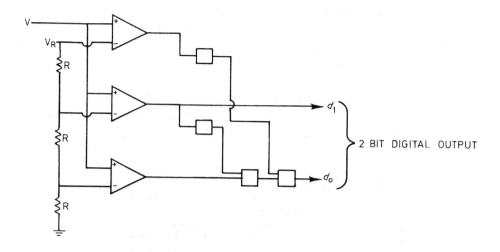

Fig 4.3.23 Two bit ADC.

Obviously the number of comparators needed becomes prohibitive,as the
resolution is increased. A more practical method is to use a DAC in closed
loop configuration. There is a further alternative,which is based on the
generation of a linear ramp voltage as a measuring scale. Three techniques
will be described, two of which are based on the first principle and one on
the second.

(a) Counter - Digital to analogue converter (Fig 4.3.24)

The DAC shown in the diagram is used to provide a voltage proportional
to the contents of a gated binary counter. To make a measurement the count
is cleared and then incremented by clock pulses until the DAC output exceeds
the input voltage V. The change of state of the comparator output signals
the completion of the conversion:

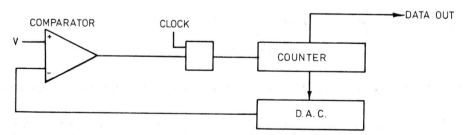

Fig 4.3.24 Counter ADC.

(b)   Successive approximation (Fig 4.3.25)

The main disadvantage of method (a) is that the conversion time can be long*.   It varies in fact from 2 to 2^N-1 clock periods for an N-bit resolution conversion.   The clock frequency itself is limited by the settling time of the DAC to about 1 MHz.

As an alternative to counting through all possible values which may correspond to the input, the ranges within which the input value lies may be examined and successively halved.   This successive approximation method is illustrated in Fig 4.3.25.

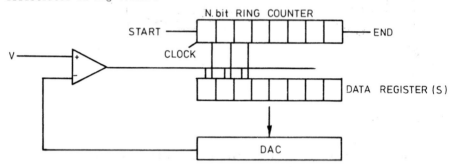

Fig 4.3.25   Successive approximation method ADC.

A ring counter of the same number of components (N) as the data register interrogates each component of the latter in turn, starting from the most significant.

Let V be the voltage to be digitized,

V_s be the trial reference voltage given by the D-A conversion of the current content of the data register S.

The conversion starts with $S = 2^{N-1}$.   The components of S are made 1 in turn starting with the most significant.   If $V < V_s$ the component is reset to 0.   Thus the first range tested is $\gtrless 2^{N-1}$.   Thereafter if

$V > 2^{N-1}$ the next value tested is $2^{N-1} + 2^{N-2}$

$V < 2^{N-1}$ the next value tested is 2^{N-2}.

When the least significant component has been set the conversion is complete.

---

*This applies only to absolute measurements.   An extension of the technique using two comparators and an UP/DOWN counter is very suitable for incremental measurements.

(c) Dual slope integration method

The method is illustrated in Fig 4.3.26.

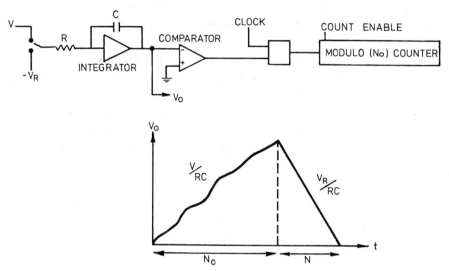

Fig 4.3.26 Dual slope integration method ADC.

The input switch is set to connect the signal V to the integrator at the same time as the counter is enabled. The integrator output voltage rises at a rate V/RC volts/s as the capacitor C charges. When the counter reaches the maximum count and returns to zero, the input switch changes over to connect an internal reference voltage - V_R to the integrator. The capacitor C discharges causing the integrator output voltage to fall at a rate of V_R/RC volts/s. The comparator senses the voltage having reached zero and disenables the counter.

If the counter is a Modulo (N_o) counter and N clock periods are counted in the second phase of the operation the result is

$$\frac{N_o V}{RC} = \frac{N V_R}{RC}$$

so that N is directly proportional to the input voltage. The merits of the technique are:

(i)   the absolute accuracy or stability of the integrating capacitor is unimportant,since both V and V_R are integrated by the same capacitor,

(ii) long term drifts in clock frequency do not affect the accuracy,since only the ratio $N:N_o$ is important,

iii   During the first phase it is the integral of the input voltage which is measured. Input noise is thus to a large extent averaged out by this process. For example, setting N_o and the clock period so that the integration takes 20 ms, would reject 50 Hz mains interference almost entirely.

To sum up; method (a) is the simplest and cheapest technique; method (b) would be used for high speed conversions and method (c) is suitable for high accuracy measurements in the presence of noise.

Fortunately it is unnecessary to build up these circuits from the component parts. A wide range of devices* are available which perform these functions. Resolutions of between 8 and 16 bits may be chosen and the output can be coded in either binary or BCD form.

ADC input voltages are usually fixed at 0-10V. Moreover ADC input impedances tend to be low. Therefore it is usually necessary to scale and buffer the signal using an amplifier between the signal source and the ADC. There are circumstances where either:

(i)   the signal to be converted is only present for a short period, or

(ii)   the object of the conversion is to measure a rapidly changing signal at a particular instant of time, or

(iii) the conversion is slow relatively to changes in the signal level.

In these cases the sample and hold circuit of Fig 4.3.27 is required. The capacitor is charged rapidly by direct connection to the signal source (V). It is then isolated and decays only very slowly because of the high input impedance of the amplifier.

For speed of operation, the switch must be solid state. It must introduce no significant additional resistance in the closed state and give good isolation in the open state. The amplifier must have a very high input impedance. (It might be of the FET variety.) A decay rate of less than 1 volt/s can then be obtained.

Here again, sample and hold circuits are available as systems blocks [4.4] .

In 4.2, where signal transmission was dealt with, it was pointed out that plant signals should always first be digitized, so that accuracy of measurement is not lost in transmission. However, under certain conditions, there are still more reliable methods of collecting plant measurements. They

*They are made as hybrid circuits and packaged in encapsulated block form [4.4].

are an alternative to ADC techniques and produce signals which are even more insensitive to transmission faults.

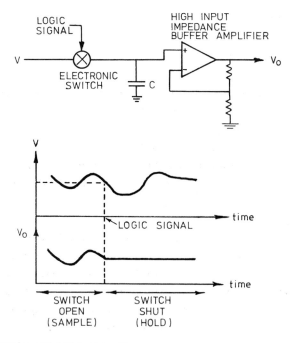

Fig 4.3.27 Sample-and-hold circuit.

One way is to use a voltage controlled oscillator (VCO) to give a train of pulses whose frequency is proportional to the input voltage.

To obtain a measurement,the controller has only to open a gate to a counting register for a preset time. Using a crystal clock in the control unit, this interval can be set extremely accurately by means of a modulo counter.

From the point of view of transmission one cable now suffices to carry the information. There is the further advantage that a single noise pulse occurring during transmission will introduce only ±1 bit error in the measurement. This contrasts with the parallel transmission of ADC data. Here the most significant bit could be corrupted and render the data useless.

The technique is suitable for ratiometric measurements with a large dynamic range, but is restricted to slowly changing signals. It is also necessary to limit the lower end of the range of input voltages, since converters cannot cope with very low frequencies. However the linearity of a voltage to frequency converter makes a 0.1% accuracy of measurement possible.

Another method of measurement, whose signals are easily transmitted, is the phase measurement described in 4.3.2d.

4.4   POWER DRIVERS

4.4.1   Electromagnetic devices

The following devices are all based on the use of a solenoid to give mechanical motion. The moving member is either

a   an electrical contact: relays and contactors,
b   coupled to the actual mechanism: solenoid actuators,
c   a valve in a pneumatic or hydraulic circuit: solenoid valves.

(a) Relays and contactors

Apart from their use in signal switching (see 4.2.2), relays are very useful interface elements in power driving. The smallest type of relay is the reed relay with typical contact rating for non–inductive loads of 1A, 250V, 15VA. With mercury–wetted relays,these ratings can be doubled or even quadrupled and for greater loads a host of contactor relays are available. Their use in heavy power switching is well established. The shortcomings of relays are

i   Contact arcing tends to erode contacts and shorten life.
ii   An element of unreliability. For continuous repetitive switching semi-conductor switches are preferred.
iii   They do not mix well with semi-conductor logic circuits since contact arcing and field changes in the solenoid are powerful sources of noise.

The manufacturers' data books give extensive technical details on the electrical and mechanical properties of relays. Recommendations are also made on ancillary components which should be added to suppress contact arcing and prevent voltage overshooting in the solenoid circuit. Many devices have protecting diodes and electrostatic shields included in the encapsulation.

(b) Solenoid actuators

The solenoid is fitted with an iron plunger which is pulled in when the solenoid is energized. The force increases as the plunger moves inwards. To give a pushing movement, an extension must be fitted to one end of the plunger. The solenoids are available in either d.c. or a.c. versions whose characteristics differ in the following ways

(i)   In an a.c. device, the inductance of the solenoid coil will decrease as the plunger is extended. The coil current therefore increases and tends to compensate for the loss of restraining force as the plunger moves more and more out of the high field region of the solenoid. d.c. devices have no such compensation and tend to be much less powerful at their full extension.

(ii) For the same reason, the current drawn by an a.c. solenoid in the extended position may be ten times greater than in the normal retracted position. This can limit the duty cycle of the operation. d.c. solenoids draw a constant current throughout the movement of the plunger.

(iii) In an a.c. solenoid the point in the mains cycle at which switching occurs, will determine the instant of operation. This gives a slight operating jitter of perhaps 10 ms. For a given load, a d.c. solenoid will have an accurately reproducible operating time.

(iv) Unless well damped, a.c. solenoids have a tendency to hum but d.c. devices are quiet.

In general, therefore, d.c. solenoid actuators are preferable for fast, light duty, short stroke applications and a.c. solenoids for heavy duty action.

The operating speed is determined by load inertia, friction, and the solenoid force over the distance travelled. The manufacturers' data books will give the necessary mechanical and electrical data to calculate the performance of a solenoid in a given application.

As a mechanical actuator, the solenoid suffers from several disadvantages.

(i)   The force varies non-linearly with distance and hence is asymmetrical when opposed by a linear spring force.

(ii) The force is limited to about 200N (50 lbf) in the most powerful a.c. solenoids

(iii) The stroke is limited to about 40 mm (1.6 in) even in a.c. solenoids. Linkages must be used to increase this.

(iv) The power dissipation in the winding resistance can easily be several hundred watts for a medium size device. As the duty cycle increases, the coil windings will heat up. The resulting increase in resistance and decrease in coil current reduces the force available from the solenoid.

(c) Solenoid valves

Miniature solenoid-operated valves make it possible to interface a wide range of pneumatic equipment to the control unit. They are suitable for driving light duty pistons and rams. The individual valves may be mounted on a manifold and the connecting air hoses attached by quick-fit couplings. They offer significant advantages over solenoid actuators. A very much longer stroke is possible—up to 150 mm (6 in). The action is linear and uniform and they generate no electrical interference.

A difficulty arises,however,with heavier loads,particularly if a long stroke is needed. At the fairly low pressures which it is practical to use, the pneumatic 'resistance' of small bore valves and hoses, restricts the speed of operation.

In these cases a stage of pneumatic amplification is necessary. This may be in the form of either

(i)   two valves; a pilot valve which is solenoid operated and provides a jet for the operation of a larger capacity valve, or
(ii)   a fluid amplifier, which incorporates the two stages in one unit. The solenoid valve closes a pilot jet. This causes pressure to build up which opens the main valve.

The first alternative is particularly useful in a hostile environment since the electrical part of the operation can be removed to a safer area.

The second is very valuable as a stage of amplification which does not generate electrical interference. Power amplification of 3 to 4 orders of magnitude are attainable by such means,with operating times of the order of 10 ms. At 5V, less than 80 mA is required to operate the solenoids. With some protection, the output stage of an integrated circuit driver can operate the solenoids directly.

The subject of hydraulic valves is vast and cannot be dealt with here. They are extensively used as servo valves for proportional movements, but simpler types are available for use as switching valves. Pressures are much higher than in pneumatic systems. Therefore much greater forces are attainable. However, oil flow is slow and operating times are longer than for pneumatic valves. The power gain is 2 to 3 times larger than with pneumatic valves, notwithstanding the much greater power required to operate the solenoids.

Hydraulic valves are operated by either d.c. or a.c. solenoids, with power consumption of the order of 20W. In this sense they are similar to the shorter stroke actuator solenoids described above. Mercury-wetted relays are a suitable interface element between DTL/TTL circuits and such solenoids. A CR filter across the contacts is needed to suppress arcing and a diode must clamp the collector of the output transistor to supply, to prevent a voltage overshoot when the solenoid is de-energized. An alternative to the relay is a semi-conductor driver of the type described in the next section. d.c. solenoids can be driven by power transistors and triacs are frequently used to drive a.c. solenoids.

4.4.2 Semi-conductor devices

(a) D.C. power control

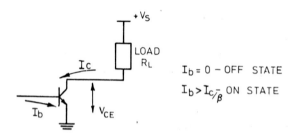

$$I_b = 0 - \text{OFF STATE}$$
$$I_b > I_c/_\beta \ \text{ON STATE}$$

Fig 4.4.1 Transistor switch.

Fig 4.4.1 shows how a transistor can be used to switch-on the d.c. current in a load. The controlling current I_b should be chosen such as to hold the driving transistor in the saturated state. The collector-emitter voltage in this state is less than 1V. The required value of collector current I_c is then known from the supply voltage V_s and load resistance R_L. The d.c. current gain β (the term forward current transfer ratio h_{FE} is also used) of the transistor is given in the manufacturers' data sheets. Hence the minimum value of I_b is established. Since β varies with temperature and is also slightly different for individual transistors, it is necessary to supply at least 2 to 3 times the minimum current to ensure that the transistor does not come out of saturation.

It is important to operate the transistor in saturation as it minimizes the power dissipation $P = V_{CE}I_c$ in the device. In the OFF state only a few μA of leakage current will normally flow, so that although $V_{CE} \cong V_s$, the dissipation is very small.

The following points should be considered in selecting a transistor switch.

(i) The V_{CE} limit, $V_{CE(max)}$

If this is exceeded avalanche breakdown can occur between the collector and the other two terminals, causing permanent damage to the transistor. For most transistors $V_{CE(max)} < 100V$, but some types can work up to 600V. The latter have rather slower switching speeds.

(ii) The I_c limit, $I_{c(max)}$

The maximum collector current is determined by the maximum power dissipation and the area of the collector junction of the transistor. The $I_{c(max)}$ ranges from a few hundred mA to 50 A depending on construction. For higher currents several transistors can be connected in parallel. Care should be taken then to ensure the current divides equally among them, by employing load sharing resistors in series with each collector.

(iii) The power dissipation limit P_{max}

This depends on the thermal resistance of the device package. High power transistors are bolted to heat sinks to give more efficient heat dissipation.

(iv) Switching speed

In power control applications, switching speeds generally matter only because of their effect on power dissipation. During switching, the product $V_{CE} I_c$ is greater than in the ON state, since the saturated state ($V_{CE} < 1V$) is only reached at the end of the switching transient. Therefore a limit on the duty cycle of a transistor switch is set by the value of the mean power, which should not exceed P_{max}.

(v) External load

Some loads may pose particular difficulties. Inductive loads, such as solenoids, must be fitted with protective diodes to prevent $V_{CE(max)}$ being exceeded when the transistor interrupts the load current. Heater or lamp filaments have a very low resistance when cold and draw large initial currents.

Discrete transistors may be driven by DTL/TTL buffer gates as shown in Fig 4.4.2a, but β is only sufficient to allow at most 100 mA to be switched. This is due to the small currents which gates are able to supply in the high state, therefore if greater currents are to be switched, an external resistor must be used to supply I_b.

Improved current driving can be obtained by using a Darlington pair of output transistors with a β equal to the product of the individual current

gains (Fig 4.4.2b). The driver stage is also turned off more firmly when the
gate output is in a low state, since the input voltage must now exceed two
base-emitter voltages ($\sim$1.2V) before conduction begins.

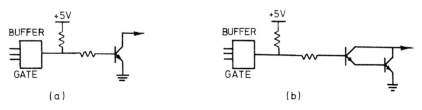

Fig 4.4.2 Discrete transistor driving circuits,(a) showing gate with enhanced
high state current supply, (b) as (a) driving Darlington pair.

(b) A.C. power control

Thyristor (SCR) and triac semiconductor switches are used to provide
'ON-OFF' or proportional control of a.c. power*. Like the transistor they
have three terminals: two power electrodes and a gate which controls the flow
of current between them.

In the thyristor (Fig 4.4.3a) the power electrodes are the anode and
cathode and conduction is unidirectional. In the triac (Fig 4.4.3b) the
electrodes are simply known as MT1 (nearer the gate) and MT2, and conduction
is bidirectional.

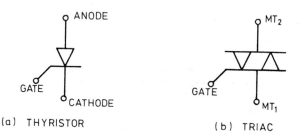

Fig 4.4.3 Semiconductor switching devices.

The thyristor may be thought of as a diode with an additional control
electrode. When it is forward biased it may be either in the OFF or ON state.

In the OFF state,conduction is blocked. Only a small leakage current
(of a few μA) will flow between anode and cathode.

---

*For 'ON-OFF' switching with low duty cycle and particularly for multiple
switching, a.c. relays may be a simpler solution. They are easily driven
from a reed relay such as described in 4.2.2 (a).

In the ON or conducting state, it behaves like a forward-biased diode. Large currents can flow and there is only a small voltage drop between anode and cathode.

The thyristor is switched on from any of the following causes.

(i) The application of a voltage between gate and cathode so as to inject a small gate current. When this gate current exceeds a threshold value, it starts a regenerative process in which the current increases to its maximum value within a few μs. Thereafter the gate has no control.

(ii) A rapidly rising voltage at the anode, which starts the conduction process.

(iii) An applied voltage greater than that which the device can support. An avalanche breakdown may then occur in the semiconductor junction.

The thyristor can only be switched off by reducing the anode current below a threshold value, after which conduction gradually ceases. The turn-off time (~50 μs) is considerably longer than the turn-on time.

When reverse-biased, the thyristor behaves like a semiconductor diode, irrespective of gate current and is effectively non-conducting.

Triacs have much the same characteristic except that there is no distinction between forward and reverse biasing. Moreover the device will trigger on both positive or negative gate currents, although it is generally more sensitive when the polarities of gate and MT2 voltages are the same.

The use of a thyristor as an a.c. switch is illustrated in Fig 4.4.4

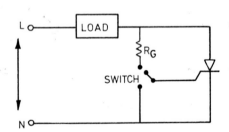

Fig 4.4.4 Thyristor switch.

When the switch is closed, gate current flows through R_G. The thyristor then conducts during each positive half cycle of the mains voltage and acts as a half wave rectifier. The point in the cycle at which conduction starts is determined by R_G. If R_G is small, the thyristor turns on at the beginning of the cycle. As R_G becomes larger the threshold gate current will not be reached until later in the cycle.

The switch may be replaced by a reed relay. The circuit may then be used in place of a high power a.c. relay. This avoids problems of contact wear due to the breaking of large currents.

Full-wave control can be obtained by using two thyristors in inverse parallel, but this requires two gate driving circuits. In most cases the use of a triac is more convenient.

The following points should be considered in designing circuits using thyristors and triacs.

(i)   Voltage and current ratings

The breakdown voltage limits the peak voltage which the device can sustain. In a reverse-biased thyristor permanent damage to the device can result from exceeding this voltage. On the other hand, the forward-biased thyristor or triac will simply switch into the conducting state and recover after the next zero-crossing of the supply voltage.

Voltage spikes on the mains are unlikely to damage the device, but can cause intermittent false triggering. For this reason it is best, in critical applications, to use a device with a higher rating than appears necessary (600V device to switch 240V a.c.). Alternatively or in addition, suppressor networks or transient clipping diodes should be used to limit the amplitude of voltage spikes.

The maximum rms current in the conducting state should be within the limits of the device chosen. The surge-current rating will however be much greater. This means that the device can be used safely to switch loads which have an initially low resistance such as stationary motors or incandescent lamps.

Fuses are recommended to protect the device against overcurrent faults. Where these are common, as for instance where a motor is liable to stall, a protective circuit breaker is preferable.

Thyristors and triacs are never completely safe from false triggering. Therefore they cannot be relied upon to give complete isolation. The inclusion of a mechanical switch or contactor in series with the semiconductor switch is a common safety measure.

(ii)   Gate characteristics

Gate currents are only needed until the principal current has been established. After the device has switched on, the gate junction is reverse-biased and no

further current flows in the gate circuit. A thyristor may therefore be
fired using a trigger pulse. This pulse must however provide a current
greater than the gate threshold current and must be of a certain minimum
duration*.

(iii) Transient conditions

     Rate of change of voltage (dV/dt)

A sudden rise in voltage across the device may cause it to breakdown into
the conducting state. Large values of dV/dt can easily arise in switching
off inductive loads. Since the applied voltage and controlled current are
in quadrature, the device switches into the non-conducting state at a time
when the applied voltage is greatest (Fig 4.4.5).

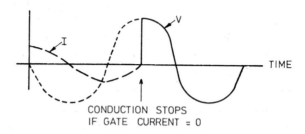

Fig 4.4.5 Current and voltage relationship when a thyristor is used to switch
off an inductive load.

    This can cause false triggering in a triac, since the full voltage is
applied immediately after conduction has stopped.

    This is one case where the use of two thyristors in inverse parallel
can be a preferable solution. Each thyristor has a quarter cycle interval
between the end of conduction and the time when the applied voltage is again
in a forward direction. This allows the device to recover and removes the
danger of false triggering.

    However, the simple addition of a CR network (Fig 4.4.6) will remove
the danger of false triggering due to this cause even in a triac.

    The rate of voltage rise is reduced by bringing the circuit into oscill-
ation with the inductance of the load. The maximum rate of change is then

$$\frac{dV}{dt} \simeq \frac{Vmax}{\sqrt{LC}}$$

---

*The manufacturers' data sheets provide the necessary information.

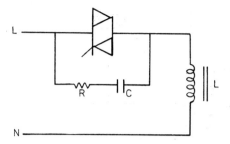

Fig 4.4.6 Protecting a triac against false triggering due to inductive loads.

This allows C to be so chosen that $(dV/dt)_{max}$ lies within the acceptable limit. A resistor R is needed to damp the resultant oscillations and a suitable value is given by

$$R \approx 2\sqrt{\frac{L}{C}}.$$

Rate of change of current (dI/dt)

Conduction in a thyristor or triac is due to a progressive breakdown of semiconductor junctions. Variations in thickness and composition of the semiconductor layers cause breakdown to occur initially at a small number of points in the junction area, before it spreads over the whole junction. For very high dI/dt local 'hot spots' may thus occur and cause premature failure of the device.

In most cases there will be sufficient inductance in the circuit to prevent this happening. Where this is not so, a choke may have to be added in the circuit.

(iv) Noise immunity

Good noise immunity is necessary for all signal lines connected to the gate since both thyristors and triacs are easily triggered by short noise pulses. The circuits of Fig 4.4.7 are commonly used to provide both noise rejection and an appropriate isolation between the control logic and a.c. drivers.

Of the three techniques (a) is the simplest. Methods (b) and (c) are more suitable for fast switching. Method (c) of course requires a pulse train as a trigger source.

Thyristor and triac switching transients generate considerable interference*.

---

*Mains filters may be necessary to prevent this being carried over and radiated by the a.c. supply.

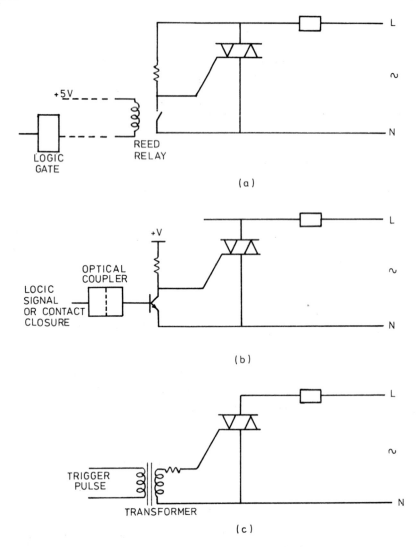

Fig 4.4.7 Coupling triggering pulses to a.c. switches.

A particularly important cause of noise, the initial switching-on of an a.c. circuit, can be minimized by using a zero voltage switching device. In this, a special circuit (Fig 4.4.8) detects the zero cross-over point of the supply voltage. The resultant pulse is gated with the control signal, so that triggering always takes place at the beginning of a mains cycle *.

---

* An optical isolator or reed relay is required for the control signal since the circuit is effectively floating.

The harmonic content of the resulting switching waveform is thereby greatly reduced since large voltage steps are avoided.

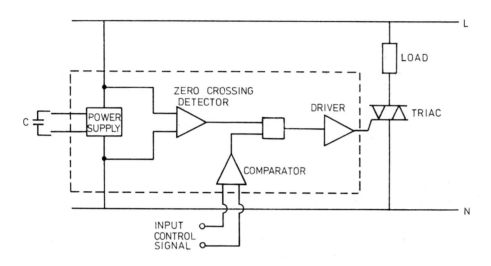

Fig 4.4.8 Zero-voltage switch.

Switching the triac at the beginning of the voltage cycle only applies if the load is resistive. With inductive loads the zero voltage cross-over point is not coincident with the zero current cross-over point. This means that a part (for purely inductive loads one half) of the supply cycle would be lost. The correct timing of the trigger pulse is just before the zero current cross-over point. This maintains continuity of conduction without introducing voltage steps.

Various types of zero cross-over switching devices are available in IC form. In some of these, the chip has its own power supply to drive the comparator and drive circuits. This is connected directly to the a.c. supply and requires only the addition of an external smoothing capacitor.

4.5   SUNDRY TECHNIQUES

4.5.1 Protection circuits

In industrial processes, it is quite common for voltage spikes to be induced in signal cables when heavy loads are being switched nearby. Some

form of protection is therefore advisable where this is likely to occur.
The circuits of Fig 4.5.1 are normally sufficient to clip low power transients.

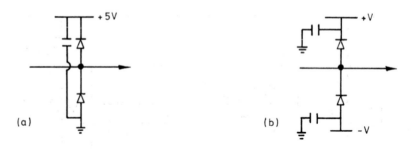

(a) For standard logic signal   (b) For bipolar signal

Fig 4.5.1 Noise-spike protection.

Other circumstances may occur where it is desirable to protect inputs
against accidental connection to a.c. power sources. Here some form of fuse
is required (Fig 4.5.2).

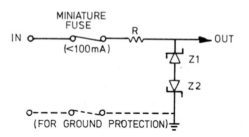

Fig 4.5.2 High voltage protection.

Zener diodes Z1 and Z2 are relied upon to limit the output voltage
during the time it takes the fuse to blow (~1 msec). The zener diodes will
be grossly overloaded during this period and must therefore be replaced at
the same time as the fuse.

The zener voltage and 'once only' dissipation (which may be a hundred
times or more its d.c. rating) determine the maximum allowable current.
This together with the mains voltage, gives the combined resistance of the
fuse (hot) and resistor R required. It should be noted that the zener voltage

will rise by a significant amount during heating (10-50 per cent).

Safety barrier circuits in flammable or explosive environments may be designed in similar fashion.

4.5.2 Driver stages for amplifiers.

Sometimes the output current rating of an operational amplifier is insufficient to drive the required load. The simple circuit of Fig 4.5.3 may be used to provide a single-ended (unipolar) output of up to 100 mA.

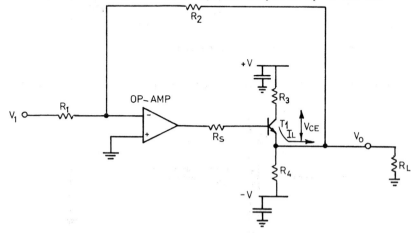

Fig 4.5.3 Transistor current amplifier to drive operational amplifier load.

Emitter follower transistor Tl acts as a current amplifier with gain β (the transistor current gain). The current supplied by the op-amp need then only be $1/\beta$ of the load current. Since the transistor current amplifier is contained within the feedback loop, there is normally sufficient negative feedback to compensate for even quite large variations in transistor charac- teristics (such as changes in β or base-emitter voltage with increased emitter current).

Resistor R_3 is normally required. It prevents excessive power dissi- pation in Tl by reducing the collector-emitter voltage (V_{CE}). It should also be chosen to keep the transistor out of saturation at the maximum rated current. R_s limits the op-amp output current and may be omitted if, as is usual, the op-amp is short-circuit protected.

To give higher output currents (up to 1 amp), two transistors Tl, T2 may be used in a Darlington pair arrangement (Fig 4.5.4) which provides higher gain.

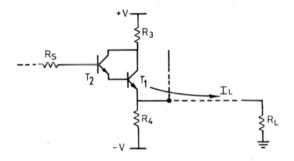

Fig 4.5.4 Circuit for higher output current.

PNP transistors may be used to provide a negative output (Fig 4.5.5).

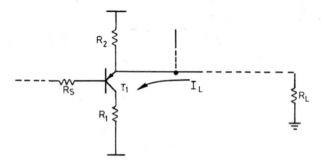

Fig 4.5.5 Circuit for negative output.

If a bipolar output is required then the best solution is to use a booster amplifier within the feedback loop. These are available,in micro-circuit form,from many manufacturers. They are less subject to drift than the above circuits and have very fast slew rates so as to preserve the orig-inal response of the op-amp.

REFERENCES

1.1   J R Ashley and A Pugh.   Logical design of control systems for sequential mechanisms, Int.J.Prod. Res. <u>6</u> No 4; 1968.

2.1   K E Iverson.   A Programming Language, John Wiley and Sons; 1962.

2.2   J H Wilkinson. Rounding Errors in Algebraic Processes, Notes on Applied Science No <u>32</u>, HMSO; 1963.

2.3   Design and Application of Microprocessors.   IEE Symposium, University of Strathclyde, Glasgow; 1974.

3.1   RCA COS/MOS Application note ICAN 6101, Data book SSD-203C; 1975.

3.2   RCA COS/MOS Application note ICAN 6267, Data book SSD-203C; 1975.

3.3   Fairchild Semiconductor, Systems Design with MSI Building Blocks, Advanced Logic Book 2; 1973.

3.4   Fairchild Semiconductor, TTL Data Book; 1972.

3.5   National Semiconductor CMOS Integrated Circuits 74C Series; 1975.

3.6   RCA COS/MOS Application note ICAN 6601, Data book SSD-203C; 1975.

4.1   IC Update Master, Publication address,645 Stewart Avenue, Garden City New York, 11530; 1976.

4.2   Teledyne Semiconductor, CMOS - HiNIL Digital Circuits; June 1974.

4.3   R Walker, CMOS specifications: 'Don't take them for granted', Electronics (McGraw Hill) Jan 9 1975.

4 4   Analog Devices Inc., Product guide; 1975.

4.5   R Parsons, Use of TTL Integrated Circuits in Industrial Noise Environments, Semiconductor Circuit Design, published by Texas Instruments Ltd, Bedford; 1972.

4.6   Fibre Optics in Communication and Data Transmission, New Electronics; Jan 21 1975.

GENERAL READING

1   Digital Electronics for Scientists, H V Malmstadt and C G Enke, W A Benjamin; 1969.

2   Introduction to Logic Circuit Theory, I Aleksander, Harrup; 1970.

3   Digital Computer Laboratory Workbook, J Hughes, published by DEC, Maynard, Mass; 1968.

4   Logic Systems Design Handbook, published by DEC, Maynard, Mass; 1972.

SUBJECT INDEX